Genetic and Anthropological Studies of Olympic Athletes

Gymnast, Caucasoid, Czechoslovakia, Age 15, Somatotype: 2½–4–3; Mean for female gymnasts: 2.7, 4.2, 2.8. (See p. 43.)

GENETIC AND ANTHROPOLOGICAL STUDIES OF OLYMPIC ATHLETES

Edited by

Alfonso L. de Garay

DEPARTAMENTO DE RADIOBIOLOGIA Y GENETICA
INSTITUTO NACIONAL DE ENERGIA NUCLEAR
MEXICO, D.F., MEXICO

Louis Levine

DEPARTMENT OF BIOLOGY
CITY COLLEGE OF NEW YORK
NEW YORK, NEW YORK

J. E. Lindsay Carter

DEPARTMENT OF PHYSICAL EDUCATION
CALIFORNIA STATE UNIVERSITY, SAN DIEGO
SAN DIEGO, CALIFORNIA

CONTRIBUTING EDITORS

Lourdes Cobo de Gallegos
George Fraser
Sipke J. Geerts
Hans Kalmus
Ching Chun Li
Ursula Mittwoch
H. Eldon Sutton

1974

ACADEMIC PRESS New York San Francisco London

A Subsidiary of Harcourt Brace Jovanovich, Publishers

The photograph on the dust jacket was kindly supplied by Arq. Pedro Ramírez Vázquez, and shows the start of the Marathon event in Mexico City, summer Olympics, 1968.

ACADEMIC PRESS, INC.
111 Fifth Avenue, New York, New York 10003

United Kingdom Edition published by
ACADEMIC PRESS, INC. (LONDON) LTD.
24/28 Oval Road, London NW1

Library of Congress Cataloging in Publication Data

de Garay, Alfonso L
Genetic and anthropological studies of Olympic athletes.

Results of a project, named the Program of Genetics and Human Biology, established through the Organizing Committee of the XIXth Olympic Games and the National Commission of Nuclear Energy of Mexico in 1967.
Includes bibliographies.
1. Athletes. 2. Anthropometry. 3. Physical anthropology. 4. Human genetics. I. Levine, Louis, Date joint author. II. Carter, John Edward Lindsay, Date joint author. III. Program of Genetics and Human Biology. IV. Title.
[DNLM: 1. Anthropology, Physical. 2. Genetics, Human. 3. Sport medicine. QT260 D317g]
GV435.G37 612'.044 74-1627
ISBN 0-12-208650-3

PRINTED IN THE UNITED STATES OF AMERICA

Contents

Lists of Participants ix
List of Research Staff xiii
Preface xv
Acknowledgments xvii

1 **Introduction** 1

I. History of the Project 1
II. Specific Approaches in This Study 3
III. General Procedures 5

2 **Physical Anthropology of the Athletes** 9

I. Previous Studies of Olympic Athletes 10
II. Methods and Procedures 14
Bibliography 24

3 **Results of Physical Anthropology Analysis** 27

I. Analysis within Sports 27
II. Analysis between Sports 50
III. Analysis by Race 61
IV. Discussion of Results 68
V. Conclusions 73

Appendix A: Instructions for Calculating the Heath–Carter Anthropometric Somatotype 74
Appendix B: Anthropometric Instruments and Measurements 76
Bibliography 82

4 Family Studies 147

I. Family Studies of Ability 147
II. Order of Birth of Athletes 152
Bibliography 155

5 Taste Thresholds for Phenylthiourea (PTC) 157

I. Introduction 157
II. Methods 159
III. Results 161
IV. Discussion 163
V. Summary 164
Bibliography 164

6 Single Gene Systems of Blood 165

I. Introduction 165
II. The Meaning of Genetic Variation 166
III. Systems Selected for Study 167
IV. Statistical Analysis for the Study of Single Gene Systems 168
V. Red Cell Antigens 169
VI. Red Cell Enzymes 172
VII. Red Cell and Plasma Proteins 178
VIII. Discussion 185
IX. Summary 185
Bibliography 186

7 Dermatoglyphic Studies: Finger and Palm Prints 189

I. Introduction 189
II. Characteristics of Dermatoglyphics 190
III. Inheritance 197
IV. Material and Methods 198
V. Results and Discussion 199
Bibliography 209

8 **Cytogenetics** 211

I. Cytological Investigation 211
II. Chromosome Analysis 219
Bibliography 224

9 **Retrospect** 229

I. Biological Approach to Sport 230
II. Implications of These Studies for Human Biology 231

Index 233

Lists of Participants

List of Participants in the First International Seminar for the Study of the Athletes of the XIXth Olympic Games, Mexico City, 17–21 July 1967

Chairman: Alfonso L. de Garay

Moderator: Louis Levine

Discussants: Douglas R. Bainbridge
Venerando Correnti
Eusebio Dávalos Hurtado
Johanna Faulhaber
Rodolfo Félix E.
Sipke J. Geerts
Santiago Genovés
Barbara H. Heath
Hans Kalmus
Ching Chun Li
Felipe Montemayor
Alejo E. Romero
Eugene Schreider
Curt Stern
Virginia Tiburcio

Advisers in Sports Medicine Attending the Seminar

Mihailo Andrejević
Gilberto Bolaños Cacho

Paul André Chailley-Bert
Aniceto Ortega
Fernand R. Plas

Secretariat

Guadalupe Estrada
Ma. Eugenia Guerrero
Rebeca Gutiérrez

List of Participants in the Second International Seminar for the Study of the Athletes of the XIXth Olympic Games, Mexico City, 25–29 March 1968

Chairman: Alfonso L. de Garay

Moderator: Louis Levine

Discussants: Douglas R. Bainbridge
Venerando Correnti
Johanna Faulhaber
Rodolfo Félix E.
Charles E. Ford
Oswaldo Frota-Pessoa
Sipke J. Geerts
Santiago Genovés
Hans Kalmus
Ching Chun Li
Rubén Lisker
Arturo Romano Pacheco
Alejo E. Romero
H. Eldon Sutton
Virginia Tiburcio
Ma. de Lourdes C. de Gallegos

Advisers in Sports Medicine Attending the Seminar

Gilberto Bolaños Cacho
Paul André Chailley-Bert
Aniceto Ortega

Secretariat

Guadalupe Estrada
Ma. Eugenia Guerrero
Rebeca Gutiérrez

List of Participants in the Third International Seminar for the Study of the Athletes of the XIXth Olympic Games, Mexico City, 16–21 June 1969

Chairman: Alfonso L. de Garay

Moderator: Louis Levine

Discussants: J. E. Lindsay Carter
Rodolfo Félix E.
Charles E. Ford
George R. Fraser
Marcel Hebbelinck
Sipke J. Geerts
Ching Chun Li
Rubén Lisker
Rosario Rodríguez
Alejo E. Romero
H. Eldon Sutton
Virginia Tiburcio
Ma. de Lourdes C. de Gallegos

Advisers in Sports Medicine Attending the Seminar

Ismael Zurita

Secretariat

Ma. Eugenia Guerrero
Rebeca Gutiérrez

List of Research Staff

Investigators:	Ma. de Lourdes C. de Gallegos
	Johanna Faulhaber
	Rosario Rodríguez
	Alejo E. Romero
	Virginia Tiburcio
Assistants:	J. Miguel Betancourt
	Guadalupe Cobo
	Aurora Chimal
	Rodrigo Diaz
	Alfonso de Garay
	Juan Manuel Herrera
	Enrique Monedero
	Olga Olvera
	Carola Rodríguez
	Mario Rios
	María Salcedo
	Mario Stoute
	María Villanueva
Technicians:	Consuelo Báez
	Martha Castañeda
	Jacques Fitch
	Graciela de Garay
	Ma. Teresa García

Guillermo Gueyde
Tania Gueyde
Carol Li
Maureen Orcutt
Rubén Ruiz

Students: Ma. José Con
Patricia Cravioto
Ismael Dávila
Welly Madinaveitia
Ma. Eugenia Ramírez
Ma. Elena Saenz
Rafael Strauss

Consultants to the Project

George W. Karp, Jr.
Dorothy Paschal

Secretariat and Administrative Staff

Rodolfo Félix E.
Ulla Britt Bjorkman
Miguel Angel García
Ma. Eugenia Guerrero
Rebeca Gutiérrez
Jirina Hlouskova
Ma. Trinidad Lahirigoyen
Andrés Sela

Hosts: Magdalena Vargas
Cristina Morales
Maricarmen Graf
Laura Domene
Marcelino Guerrero
Javier Moreno Valle
Rubén Bielak

Preface

The Olympic Games offer unique opportunities to study various aspects of the biology of athletes, especially their genetic and anthropological characteristics. Some findings and conclusions of such studies have applications in the fields of human biology, physical education, sport medicine, and may indeed benefit all humanity by providing a better understanding of human excellence and diversity.

This study of Olympic athletes analyzes many characteristics of athletes of both sexes, from all parts of the world, and belonging to all racial groups. It is evident that athletes constitute a heterogeneous population composed of individuals in whom certain hereditary material exists which permits them to respond with a particularly outstanding sports performance to the environmental factors involved in training. It is also evident that those traits involved in athletic ability are found in all human beings at various levels of excellence. It was not surprising, therefore, that the project to investigate the athletes of the XIXth Summer Olympic Games in Mexico City aroused wide interest and that 1265 athletes from 92 countries volunteered to cooperate. In addition, about 370 nonathletes from Mexico volunteered to serve as a reference group.

All the investigations conducted in this study of Olympic athletes can be grouped into three main categories: (*1*) *family studies,* (*2*) *anthropological analyses,* and (*3*) *genetic characterizations.* The family studies were intended to reveal any possible roles of parents and siblings in the development of the athlete's career. A most unexpected and surprising discovery was the apparent importance of "birth order" of the Olympic athletes. The anthropological investigations involved the gathering of the most comprehensive series of body measurements to date on so large a number of outstanding athletes. This has resulted in the most complete description yet available of the physique of the various athletic types. Included in the anthropological study are detailed analyses of those important body components which overlap in the different sport specialties. With respect to physique, it was most interesting to find that Olympic champions do not possess the extreme body types that might be expected for their particular sport specialties, although, to be sure, they are close to it. The genetic investigations included studies of chromosomes, blood groups, blood proteins, taste sensitivity to PTC, and finger and palm prints. It was found that for certain genetic traits the Olympic-athlete population appears to differ significantly from the racial population to which it belongs. An exciting discovery was the presence of sickle-cell gene carriers among the Olympic athletes.

As indicated in the text, photographs and somatocharts of many of the participating athletes were made. Inasmuch as it was not feasible to obtain legal releases from the athletes for the use of these in publication, they had to be omitted.

It is hoped that the material presented in this volume will provide a better understanding of the genetic and anthropological characteristics that contribute to the development of athletic excellence than heretofore has been available. The information should find application in physical education programs, early identification and selection of potential athletic types, and, perhaps most important, a deeper and more significant understanding of the nature of human diversity.

Alfonso L. de Garay
Louis Levine
J. E. Lindsay Carter

Acknowledgments

This study was made possible by the liberal and wholehearted support of the *Organizing Committee of the XIXth Olympic Games* whose chairman, Arq. Pedro Ramírez Vázquez, included the study of genetics and anthropology of the athletes in the cultural program of the games; and of the *National Commission of Nuclear Energy of Mexico* whose chairman, Mr. José Gorostiza, made the facilities of the Program of Genetics and Radiobiology available.

We would also like to thank *Syntex, S. A.*, for its generous grant, which was most helpful in the processing of the data and the preparation of this manuscript for publication; and the *Hyland Laboratories* of Los Angeles, California, for their generous gift of the antisera used in this investigation.

1

Introduction

I. History of the Project

This project, named the *Program of Genetics and Human Biology*, was established through the cooperative efforts of the Organizing Committee of the XIXth Olympic Games, whose chairman was Arq. Pedro Ramírez Vázquez, and the National Commission of Nuclear Energy of Mexico, whose president was Mr. José Gorostiza. In addition, the International Olympic Committee, under the presidency of Mr. Avery Brundage, and the International Federation of Sport Medicine, under the presidency of Professor Paul André Chailley-Bert, gave their wholehearted support.

The Program of Genetics and Human Biology was established and its activities were begun as of February 1, 1967. Initially all work was done by the geneticists, technicians, and administrative staff of the National Commission of Nuclear Energy of Mexico in the facilities and with the equipment of that organization. Subsequently, the Organizing Committee of the Olympic Games voted the necessary funds for the project and included it among the 20 events of the Cultural Program of the Olympics, which was under the direction of Mr. Oscar Urrutia.

This aid made possible the expanded development of the program, which involved a considerable increase in the technical and administrative staff. The collaboration of physical anthropologists, anthropology students, and laboratory technicians was enlisted; foreign specialists in human genetics and physical anthropology were also invited to Mexico for varying periods of time.

The Organizing Committee also financed preparatory meetings in which Mexican and foreign scientists participated. The First International Seminar for the Study of the Olympic Athletes took place in July, 1967, in the Sala de Comisiones del Comité Permanente Interamericano de Seguridad Social in Mexico City. During this seminar the lines of research were tentatively defined, and the genetic tests and anthropological measurements that were to comprise the investigation were selected.

Preliminary studies were conducted during the IIIrd International Sports Competition in October, 1967, when 196 athletes, both male and female, participated in the various tests. For this initial phase of the project an additional laboratory was set up next to the Genetics Laboratory of the National Commission of Nuclear Energy, mainly for the anthropological examination of the athletes who participated in the IIIrd International Sports Competition and for a reference group of about 370 young men and women.

In March, 1968, the Second International Seminar for the Study of the Olympic Athletes took place. The national and foreign investigators evaluated the research carried out during the IIIrd International Sports Competition and determined the lines of research for the Olympic Games. After each of the first two International Seminars a report was published, outlining the work of the respective seminar. In September, 1968, the Organizing Committee made available a building in the Olympic Village and the Genetics and Human Biology Laboratory was set up there for the examination of the Olympic athletes. This laboratory worked steadily during September and October, both with the athletes and the reference group.

Before the Games, Arq. Ramírez Vázquez wrote to the presidents of all the National Olympic Committees asking for their cooperation in enlisting their athletes' participation in the project. The letters were well received, and most National Committees recommended participation to their delegations. At the beginning of November, 1968, the examinations in the Olympic Village laboratory came to an end and it was closed down. The analysis of the results proceeded in the laboratories of the Program of Genetics and Radiobiology of the National Commission of Nuclear Energy in Mexico City. The raw data were encoded, transferred to punch cards, and the processed data distributed for analysis. Subsequently, the manuscripts dealing with the various genetic and anthropological aspects of the study were collected and edited.

During the initial stage of the analysis of the results and the preparation of the manuscripts, the program with the help of Arq. Ramírez Vázquez received a supplementary grant from Syntex, S. A. The Third International Seminar took place in June, 1969, under the auspices of the National Commission of Nuclear Energy and with the support of Syntex, S. A. Subsequent to the completion of this Seminar, editorial work on the project continued with the collaboration of the staff of the Genetics Program of the National Commission of Nuclear Energy and of invited experts.

II. Specific Approaches in This Study

In this study we are interested in obtaining a body of knowledge, through a meticulous analysis of the athletes' genetic and anthropological characteristics, that can be applied extensively to human well-being. With this principle in mind, the following approaches were established:

1. The investigation of the sports history of the families of the athletes.
2. The investigation of a number of genetic traits of the athletes, especially those that might be correlated with sports ability.
3. The investigation of those aspects of the athletes' physiques that represent the interaction of their genetic endowments with the environmental factors involved in training.
4. The application of these investigations to human genetics and biology.
5. The application of these findings to possible indications, even in early childhood, of the sports speciality for which an individual is best suited.

The present study of the Olympic athletes provided an opportunity to study more than 1000 athletes of both sexes, from all parts of the world, and belonging to all racial groups. The application of our findings to the well-being of all human beings rests on the fact that the traits that contribute to athletic ability are found in all human beings in varying measure.

TABLE 1.1. PARTICIPANTS IN THE XIX OLYMPIC GAMES BY COUNTRY AND NUMBER OF ATHLETES STUDIED

COUNTRY	PARTICIPANTS			LISTED FOR STUDY	ACTUALLY STUDIED			
	M	F	TOTAL		M	F	TOTAL	% OF TOTAL PARTICIPANTS
AFGHANISTAN	6	0	6	6	4	0	4	66.66
ALGERIA	3	0	3	3	3	0	3	100.00
ARGENTINA	87	5	92	44	26	0	26	28.26
AUSTRALIA	112	24	136	93	17	1	18	13.23
AUSTRIA	36	8	44	31	2	0	2	4.54
BAHAMAS	20	0	20	12	7	0	7	35.00
BARBADOS	9	1	10	9	5	0	5	50.00
BELGIUM	83	5	88	48	17	3	20	22.73
BERMUDA	9	0	9	2	2	0	2	22.22
BOLIVIA	5	0	5	1	1	0	1	20.00
BRAZIL	81	4	85	37	24	2	26	30.69
BRITISH HONDURAS	7	0	7	5	0	0	0	0.00
BULGARIA	111	11	122	82	14	0	14	11.47
BURMA	4	0	4	4	0	0	0	0.00
CAMEROUN	7	0	7	7	3	0	3	42.86
CANADA	117	28	145	105	44	8	52	35.86
CENTRAL AFRICA	1	0	1	1	1	0	1	100.00
CEYLON	3	0	3	2	2	0	2	66.66
COLOMBIA	43	5	48	29	8	0	8	16.66
CONGO	1	0	1	1	0	0	0	0.00
CONGO KINASHA	10	0	10	10	5	0	5	50.00
COSTA RICA	19	2	21	15	10	1	11	52.38
CUBA	108	16	124	103	36	9	45	36.29
CZECHOSLOVAKIA	96	28	124	74	28	7	35	28.22
CHAD	4	0	4	4	1	0	1	25.00
CHILE	21	2	23	14	8	2	10	43.48
DENMARK	68	4	72	52	10	0	10	13.89
DOMINICAN REPUBLIC	23	0	23	19	15	0	15	65.24
EAST GERMANY	212	41	253	190	12	0	12	4.74
ECUADOR	15	1	16	16	11	2	13	81.25
EL SALVADOR	57	8	65	33	13	8	21	32.31
ETHIOPIA	21	0	21	21	8	0	8	38.10
FIJI (ISLANDS)	1	0	1	1	1	0	1	100.00
FINLAND	64	6	70	57	5	1	6	8.57
FRANCE	178	32	210	121	0	0	0	0.00
GABON	1	0	1	1	1	0	1	100.00
GERMANY	251	45	296	220	7	0	7	2.36
GHANA	36	1	37	18	15	0	15	40.54
GREAT BRITAIN	186	51	237	165	13	3	16	6.75
GREECE	48	0	48	33	30	0	30	62.50
GUATEMALA	52	1	53	24	23	1	24	45.28
GUIANA	5	0	5	5	0	0	0	0.00
GUINEA	18	0	18	0	0	0	0	0.00
HONDURAS	10	0	10	10	10	0	10	100.00
HONG-KONG	12	0	12	3	0	0	0	0.00
HUNGARY	152	38	190	141	14	2	16	8.42
INDIA	28	0	28	8	2	0	2	7.14
INDONESIA	13	0	13	7	5	0	5	38.46
IRAN	14	0	14	14	3	0	3	21.43
IRAQ	3	0	3	3	1	0	1	33.33
ICELAND	6	2	8	8	1	0	1	12.50
IRELAND	27	7	34	18	6	0	6	17.65
ISRAEL	28	3	31	8	0	2	2	6.45
ITALY	184	17	201	150	32	1	33	16.42
IVORY COAST	12	0	12	12	11	0	11	91.67
JAMAICA	29	6	35	29	12	0	12	34.28
JAPAN	158	29	187	110	25	0	25	13.37
KENYA	40	3	43	22	1	1	2	4.65
KOREA	44	14	58	44	2	0	2	3.45
KUWAIT	2	0	2	2	0	0	0	0.00
LEBANON	12	0	12	5	0	0	0	0.00
LIBYA	1	0	1	1	1	0	1	100.00
LIECHTENSTEIN	2	0	2	2	2	0	2	100.00
LUXEMBOURG	3	2	5	3	1	0	1	20.00
MADAGASCAR	4	0	4	4	0	0	0	0.00
MALAYSIA	33	0	33	15	7	0	7	21.21
MALI	2	0	2	2	2	0	2	100.00
MALTA	1	0	1	0	0	0	0	0.00
MOROCCO	26	0	26	26	25	0	25	96.15
MEXICO	253	47	300	189	143	19	162	54.00
MONACO	2	0	2	0	0	0	0	0.00
MONGOLIA	16	4	20	17	0	0	0	0.00
NETHERLANDS	96	27	123	96	18	2	20	16.26
NETHERLANDS ANTILLES	3	2	5	2	0	0	0	0.00
NEW ZEALAND	54	5	59	35	18	5	23	38.98
NICARAGUA	13	0	13	13	7	0	7	53.85
NIGER	2	0	2	2	2	0	2	100.00
NIGERIA	36	6	42	23	19	6	25	59.52
NORTH KOREA	26	0	26	22	0	0	0	0.00

TABLE 1.1. (CONTINUED)

COUNTRY	PARTICIPANTS M	PARTICIPANTS F	PARTICIPANTS TOTAL	LISTED FOR STUDY	ACTUALLY STUDIED M	ACTUALLY STUDIED F	ACTUALLY STUDIED TOTAL	% OF TOTAL PARTICIPANTS
NORWAY	41	9	50	30	9	0	9	18.00
PAKISTAN	20	0	20	2	2	0	2	10.00
PANAMA	17	0	17	17	2	0	2	11.76
PARAGUAY	1	0	1	0	0	0	0	0.00
PERU	16	14	30	13	7	0	7	23.33
PHILIPPINES	46	4	50	40	24	4	28	56.00
POLAND	148	38	186	123	45	0	45	24.19
PORTUGAL	22	2	24	9	6	2	8	33.33
PUERTO RICO	58	4	62	44	8	0	8	12.90
RUMANIA	69	17	86	64	7	4	11	12.79
SAN MARINO	4	0	4	2	0	0	0	0.00
SENEGAL	22	0	22	22	22	0	22	100.00
SIERRA LEONE	3	0	3	3	2	0	2	66.66
SINGAPORE	4	1	5	4	0	0	0	0.00
SPAIN	127	2	129	70	21	1	22	17.05
SUDAN	5	0	5	5	5	0	5	100.00
SURINAM	1	0	1	1	0	0	0	0.00
SWEDEN	94	16	110	79	24	4	28	25.45
SWITZERLAND	88	5	93	62	13	2	15	16.13
SYRIA	3	0	3	3	0	0	0	0.00
TAIWAN	35	8	43	32	16	3	19	44.19
TANZANIA	4	0	4	4	0	0	0	0.00
THAILAND	42	0	42	12	11	0	11	26.19
TRINIDAD AND TOBAGO	20	0	20	18	8	0	8	40.00
TUNISIA	7	0	7	7	3	0	3	42.86
TURKEY	29	0	29	26	12	0	12	41.38
UGANDA	11	0	11	11	1	0	1	9.09
UNITED ARAB REPUBLIC	31	0	31	24	1	0	1	3.22
U.S.A.	294	95	389	307	26	31	57	14.65
URUGUAY	22	6	28	23	4	6	10	35.71
U.S.S.R.	259	70	329	251	12	0	12	3.65
VENEZUELA	27	0	27	15	12	2	14	51.85
VIET-NAM	8	2	10	6	2	0	2	20.00
VIRGIN ISLANDS	8	0	8	3	0	0	0	0.00
YUGOSLAVIA	61	10	71	65	10	3	13	18.31
ZAMBIA	7	0	7	7	0	0	0	0.00
TOTAL	5240	844	6084*	4168	1117	148	1265	20.79

* Data obtained from the Official Report entitled "Participants of the XIX Olympic Games". Comité Olímpico Mexicano. México, 1968.

III. General Procedures

A. *The Population*

Olympic competitions permit the study of a fraction of the world population, of both sexes, having their origins in different regions of the globe. The individuals studied in this program belonged to 92 out of the 115 delegations represented at the games. These individuals comprised 30% of the total number of competitors in those sport specialties chosen for this study (Table 1.1).

B. *The Sample*

The athletes participating in the Olympic Games represent a particularly favorable group for the study of their biological characteristics, because they have been subjected to a selection process through the numerous competitions, tending to identify the ones who are best fitted for each sport. Such selection, made with the sole purpose of finding in each locality those individuals who are best endowed for success, favors biological research. The Olympic Games include 19 different sports specialities, each one formed by subspecialities (events that give greater heterogeneity to the sample). However, only athletes from 12 sports were chosen for measurements. The sports chosen (see Section III,C) were those in which it was felt that there existed the greatest possibility of discovering both anthropological and genetic traits that were correlated with the sport. In addition to sports and events, sex and race were categories which were selected as bases for statistical analyses (Table 1.2).

TABLE 1.2. OLYMPIC ATHLETES BY SEX AND RACE

Race				M	F	Total
Caucasoid				631	87	718
Negroid	M	F				
(a) Negroid	156	24				
(b) Mulatto	42	6				
(c) Negroid + Mongoloid	2	0				
(d) Mestizo + Negroid	4	1				
			Total	204	31	235
Mongoloid				84	3	87
Mestizo				194	27	221
Unknown				4	0	4
TOTAL				1117	148	1265

C. *Grouping of Sports*

In some cases an analysis of the genetic and anthropological data required a grouping of various sport specialities; the choice of grouping was based upon a consideration of the dominant characteristic of each type of event. These groups are: (*1*) *explosive* events: weight lifting, throwing, boxing, and jumping; (*2*) *speed* events: short-distance running, swimming, and certain cycling events; (*3*) *skill* events; gymnastics, diving, pole vaulting, pentathlon, decathlon, modern pentathlon, basketball, and wrestling; and (*4*) *endurance* events: long-distance running, swimming, certain cycling events, water polo, rowing, and canoeing.

D. *Implementation*

The sample includes athletes belonging to the specialities that are listed in Chapter 3. These specialties were determined above all by the apparent relationship that exists between the sports specialty and the physical structure of the individual. The sports specialities in which there appear to be large physical variability among the athletes and little obvious relationship between sports efficiency and physical characteristics, as measured by the anthropological tests used in this study, were eliminated. This does not, of course, preclude the possibility that sports such as pistol shooting and sailing require certain outstanding physical characteristics. Their elimination from the present study merely indicates that they could not be studied efficiently under the circumstances. Equestrian sports were eliminated due to the large extent to which the animal affects the outcome of the competition.

The project examined 1265 Olympic athletes from the total number of 6084 competitors. This represents some 30% of the 4168 athletes involved in the 12 sports designated for this study. Of the 110 countries (from the total of 115) that had one or more athletes scheduled to compete in at least one of the designated events, 92 participated in this study. The examination period extended from October 1, 1967, through October 29, 1968.

Under the designation of public relations are placed all activities that function to induce athletes to participate in the Project. These include advertising the aims, purposes, and procedures of the Project as well as facilitating and rewarding athlete participation. These are two main groups of people whose approval and cooperation must be obtained in order to achieve sufficient participation of the athletes. The first group consists of the chiefs of delegations and the team physicians. The second group includes the team of trainers (coaches) and the athletes themselves. Approval of the project by the chiefs of delegations and team physicians is a necessary but by no means sufficient condition for obtaining athlete participation. The situation is further complicated by the fact that a very different type of approach is required in enlisting the cooperation of the members of the two groups.

Chiefs of delegations and team physicians are basically interested in the long-range, scientific value of the project. They are generally amenable to the idea of using the opportunity presented by the Olympic Games for a study of human achievement, including its genetic and anthropological aspects. In somewhat sharp contrast, the team trainers and the athletes themselves tend to be uninterested in long-range scientific projects. They are intensely interested in the practical aspects of any investigation and are rather hesitant to participate in any effort that is not directly involved in preparing them for the competitions. As can be readily appreciated it was found that the chiefs of delegations and team physicians are best approached by a scientific expert while the team trainers and the athletes themselves are best approached by a sports enthusiast.

2

Physical Anthropology of the Athletes

The morphological characteristics of man at work and at play have often received the attention of the human biologist or anthropologist. However, the increasing attention to sport as a prominent form of human behavior in many societies has posed many unanswered questions. The Olympic Games present a situation which is potentially very valuable for anthropological study. One may consider the human body as a system which has certain functions to perform and which has certain basic materials available to it: fat, viscera, musculoskeletal tissue, the senses, and the nervous system. We would reason that the amounts of each material in any physique are largely determined genetically, although besides genetic variation, some modification can occur, as is found for example when fat is reduced during strenuous training, or when muscles are hypertrophied in weight training. Precisely how the components or materials of physique are distributed in different parts of the body seems likely to depend upon two factors: first, the amount of each component present; and second, the nature and order of priority of the functions or tasks which the organism has to carry out. Studies on both animal and human populations give credence to this concept. Moreover, if we accept the biological dictum that there is a relationship between structure and function, then by defining tasks an estimate of the appropriate structure for optimum performance of the task may be attempted.

In most populations, one can make only hypotheses regarding what the tasks may be, and what their order of importance is, and one has little or no way of verifying whether one's postulates are correct. The importance of the Olympic Games for our purpose is twofold. First, the participants represent the elite in their chosen task; second, the tasks or sports are standardized. As anthropologists, we are provided with a set of experimental situations in which the priority of, or emphasis on, different functions changes. Thus, by examining the association between structure and functions we are able to define both aspects more clearly.

I. Previous Studies of Olympic Athletes

Although there have been many studies on athletes at other levels of competence, there are only a few which report data on the body structure of the Olympic athlete. These latter studies range from the reporting of age, height, and weight, to more extensive studies containing many anthropometric measurements, somatotyping, and tissue analysis. Several studies are briefly reviewed below because of their relevance to the present project.

Kohlrausch (1929) reported on measurements made on approximately 300 athletes at the 1928 Olympic Games in Amsterdam. (Statistical analysis was not made on some 30 women and 20 male gymnasts because of incomplete data.) Mean values were presented for 104 track and field competitors, 18 boxers, 8 wrestlers, 18 weight lifters, 14 water-polo players, 25 rowers, 23 cyclists, and an unspecified number of marathon runners and swimmers. In addition to measurements of height, weight, vital capacity, thorax circumference, upper arm circumference, and calf circumference, a small number of ratios were calculated, and photographs were presented for 53 athletes. Eighteen countries were represented, the majority from Europe, while Canada, Japan, India, Mexico, and Chile represented the rest of the world. Although the statistical analysis was limited, the data indicated that there were differences in body dimensions between events, and the first adequate data on the body size of Olympic athletes in different sports were presented. Within sport specialities, size differences were noted, and these were apparently related to the stature of the people of the country of origin. Furthermore, Kohlrausch compared the Olympic athletes with 307 successful athletes from the same sports and observed similar trends as to their physical characteristics.

As part of a study of physical fitness of champion athletes, Cureton (1951) studied the structural and functional capacities of 21 male swimmers who were members of the 1948 United States Olympic team or were of "championship ability," and 24 track and field champions who were either contenders for the United States Olympic team or championship performers at the national level. Both groups contained several Olympic champions. A considerable number of lengths, circumferences, and diameters were taken, as well as estimates of body fat, and somatotype ratings which were adaptations of Sheldon *et al.* (1940) ratings by Cureton's method. Cureton was the first to demonstrate differences in somatotype distribution among events. In addition, his data indicated that structural and compositional differences could be related to the sports speciality.

Correnti and Zauli (1964) measured 166 Olympic track and field competitors and eight swimmers at the 1960 Rome Olympics. Forty-five characteristics were analyzed and studied in the following groups: cephalic characteristics, lengths, diameters, circumferences, ratios, and skinfolds. In addition, four body photographs and two head photographs were taken. Thirty-four of the athletes were medal winners. The athletes were from 38 countries, with 46% being from the British Commonwealth and the United States, while Italy, Germany, and Russia made up another 21%. The majority of the athletes were Caucasoids, with Negroids, Mestizos, and Mongoloids in small numbers. Comparisons were made between the various track and field events, and values were compared with those of 231 third-year physical education students from the Istituto Superiore di Educazione Fisica di Roma. The most important findings may be generalized as follows:

1. Differences were observed in age, height, and weight among the various events.
2. Within certain events body shape or form was similar but size varied.
3. Relationships were observed between body proportions, dimensions, weight with proportions, and dimensions and performance.
4. Perimetric asymmetry of limbs was a distinguishing characteristic in some sport specialities.
5. Skinfold values, although relatively low for most events, were considered useful indicators of the degree of training.
6. When selected dimensions were compared with the data of Woringer (1927), an increase in dimensions but similarity in proportion was noted.

Tanner (1964) also studied track and field athletes at the Rome Olympics. One hundred and thirty-seven track and field athletes from 23 countries were measured. The majority of the athletes (73%) were from the British Commonwealth and the United States, with African, European, and Asian countries poorly represented. Of the 137 subjects, 113 were Caucasoid, 15 Negroid, and 9 Mongoloid.

Tanner's analysis was presented in terms of differences in somatotype, size, proportions, and tissue analysis. A unique contribution was made by the use of discriminant function analysis. The most important findings were as follows:

1. Caucasoid and Negroid athletes differed significantly on selected variables within events; however, the generalized findings applied equally well to both groups among events.
2. The somatotype distribution of the athletes was very different from that of the general population.
3. There were marked differences in somatotype distributions among competitors in different events.
4. In terms of body size, lengths, circumferences, proportions, and amounts of tissues in limbs, there were outstanding differences among the means of different events; however, there was some overlapping of individuals between events.
5. Certain dimensions discriminate athletes by event more clearly than do others.
6. When selected data from the 1960 Olympic Games athletes were compared with the 1928 data of Kohlraush, the same general differences among events were observed and a considerable increase in size also noted.

At the 1964 Olympic Games in Tokyo, Hirata (1966) studied the size and age of the competitors. Considerable differences were observed in age, height, and weight of participants in different events. He presented extensive data on differences among countries in regard to the body size of their people, and he suggested that countries with people whose general physique was limited to the characteristics of champions in certain events should concentrate on those events.

Age as a factor of some importance in athletic performance was carefully examined by Jokl (1964) using the data from the 1952 Helsinki Olympics collected by Karvonen and others. He found that the span of years between ages of the youngest and the oldest competitors was more than four decades. He also observed characteristic and significant age trends for various sports, although there were individual exceptions.

In an extensive monograph, Tittel (1965) made a comparative study on the bodily structure of the best athletes in 100-meter track (men and women), high jump (men), broad jump (men), discus throwing (men and women), and shot put (men), based upon measurements taken on the best competitors in the German Democratic Republic on the one hand, and data from athletes competing in corresponding events at the Olympic Games in Rome and Tokyo, on the other hand. By using single and multiple regression techniques, he found mathematical expressions for determining the degree of relationship between certain body measurements and athletic performance.

Finally, Azuma (1964) edited a report with tabulated data based only on the files of the Olympic Medical Archives, including height, weight, somatotype ratings, and chest girth measurements. The examination forms were completed in the home country of the respective competitors of the Tokyo Olympic Games. A total of 1110 athletes representing 23 nations were reported, of which more than half were from Asian countries, racially Mongoloids.

From the review of literature, the importance of certain parameters can be assessed. For the present study it was felt important to select for further examination parameters that had shown promise in the past, and to add new techniques that might prove of value in the future. In this manner, comparisons could be made with former work and new insights might be gained into the structural relationships of Olympic Games competitors. By measuring appropriate dimensions such as weight, height, lengths, diameters, circumferences, and skinfolds, as well as taking a somatotype photograph, analysis by size, shape, composition, and proportion can be made in relation to the sport.

While the reviewed studies have provided us with much useful information, those studies of Kohlrausch (1929), Correnti and Zauli (1964), Tanner (1964), and Tittel (1965) have certain important limitations. The emphasis in these studies has been on track and field athletes, but when these are analyzed by event the numbers have been as low as two in some categories. That certain distinct differences have already been observed is encouraging, but greater numbers would certainly give us more confidence in the results. Furthermore, the data of Correnti and Zauli (1964) and Tanner (1964) overlap considerably because 79 athletes were included in each study, although they were measured independently. Certain countries and races have not participated in the Olympics to any great extent, but with each succeeding Olympics the number has increased. On the other hand, some countries which have had relatively large delegations have not been well represented in the studies.

There are 19 sports included in the modern Summer Olympics, but apart from track and field, little is known of the differences between Olympic competitors in the different sports. In addition, women have competed in the Olympics since 1900, however, except for age, height, and weight, almost no data of an anthropometric nature has been obtained. At the 1968 Games there were six sports which had separate categories for women.

A further limitation to the previous studies is that the approaches were confined to the morphological aspects and performance. Very few studies have attempted to examine the possible genetic bases for structure and sport ability, although Tanner (1964) makes the pertinent observation that athletes are both born and made. While it is recognized that environment can influence composition and shape of the body (Lasker, 1947, Keys *et al.*, 1950), it appears that genetics prescribes the limits of such variations, although the precise mechanisms are as yet unknown. After studying 351 pairs of twins, Gedda (1961) concluded that sports activity, besides exogenous influences such as social environment, training, economic levels, etc., is certainly also conditioned by hereditary factors. As some biological characteristics of the human body have been established as solely hereditary, it might be supposed that certain morphological characteristics related to sports ability might also be identified. If such relationships were to be established, then athletes could be described as a continuum embracing genetics, structure, and function.

The authors cited are not to be condemned for the limitations in their studies, but rather to be praised for their important contributions. It is the responsibility of later investigators to improve on previous work. It is hoped that the present study will make an important contribution, not only as far as descriptive morphology is concerned, but also in terms of ascertaining the genetic differences which might be related to structure and function.

II. Methods and Procedures

A. *Physical Anthropology*

The data collected concerned two complementary aspects: (1) the somatotype, where the physique of the athlete is considered as a whole; and (2) somatometry, where anthropometric measurements are considered singly, or in combination.

1. *Somatotyping*

The assessment of somatotype was first described by Sheldon *et al.* (1940). Since that time a number of modifications have been made by other investigators. One of the modifications—recently described by Heath and Carter (1967)—is used as the major classification in this study.

Heath and Carter (1967) define the somatotype as "a description of present morphological conformation. It is expressed in a three-numeral rating, consisting of three sequential numerals, always recorded in the same order. Each number represents evaluation of one of the primary components of physique which describes individual variations in human morphology and composition [p. 70]."

In this system the first component, endomorphy, refers to the relative fatness and leanness in individual physiques. The second component, mesomorphy, refers to the musculoskeletal development per unit of height and can be thought of as relative lean body mass. The third component, ectomorphy, refers to the relative linearity of individuals. These last ratings are based largely on height–weight ratios (HWR; height/$\sqrt[3]{\text{weight}}$), so that the low ends of their distributions both connote relative shortness of the several body segments, and the high ends connote elongation or linearity of body segments. The ratings on each component begin in practice at one half and have no arbitrary end point. There are three ways of obtaining the somatotype rating. First, by means of physical anthropometry; second by means of a somatotype photograph; and third, by combining the two procedures. The rating method chosen for this study is the Heath–Carter anthropometric somatotype. The selection of this method was based on the following points.

1. The greatest amount of data on the athletes was available in this system (photographs were not available for 7% of the subjects).
2. The anthropometric procedure is easily duplicated by other investigators and does not require a photograph.
3. Preliminary data from the IIIrd International Sports Competition in 1967 indicated that better discrimination was possible on the component scales (especially mesomorphy) than on the Sheldon (1965) or Parnell (1958) scales.
4. The method can be used for males and females and corrections for age are avoided.
5. Considerable data on reference populations are available in Heath–Carter ratings.

The H–C AS is objectively obtained as follows: (*a*) by using the sum of triceps, subscapular, and suprailiac skinfolds, the endomorphy is determined; (*b*) biepicondylar diameters of humerus and femur, flexed arm girth (corrected for triceps skinfold), calf girth (corrected for calf skinfold), and height are used to determine mesomorphy; (*c*) the HWR is used to determine ectomorphy. An example of this procedure using the mean data for basketball players is given in Figure 2.1. Appendix A in Chapter 3 includes instructions for calculating the Heath–Carter anthropometric somatotype.

HEATH-CARTER SOMATOTYPE RATING FORM

NAME Olympic Games Basketball Players AGE 24.0 SEX: **M** F NO: 63

OCCUPATION Various COUNTRY Various DATE October 1968

PROJECT: Olympic Games MEASURED BY: C.R.Z.

Skinfolds (mm):
Triceps = 6·1
Subcapular = 8·6
Supraliac = 6·5
TOTAL SKINFOLDS = 21·2
Calf = 5·2

TOTAL SKINFOLDS (mm)																								
Upper Limit	10.9	14.9	18.9	22.9	26.9	31.2	35.8	40.7	46.2	52.2	58.7	65.7	73.2	81.2	89.7	98.9	108.9	119.7	131.2	143.7	157.2	171.9	187.9	204.0
Mid-point	9.0	13.0	17.0	**21.0**	25.0	29.0	33.5	38.0	43.5	49.0	55.5	62.0	69.5	77.0	85.5	94.0	104.0	114.0	125.5	137.0	150.5	164.0	180.0	196.0
Lower Limit	7.0	11.0	15.0	19.0	23.0	27.0	31.3	35.9	40.8	46.3	52.3	58.8	65.8	73.3	81.3	89.8	99.0	109.0	119.8	131.3	143.8	157.3	172.0	188.0
FIRST COMPONENT	½	1	1½	**2**	2½	3	3½	4	4½	5	5½	6	6½	7	7½	8	8½	9	9½	10	10½	11	11½	12

Height (in.) = 74·4
Bone: Humerus (cm) = 7·4
Femur = 10·2
Muscle: Biceps (cm) 32·6 − ·6 = 32·0
− (triceps skinfold)
Calf 38·3 − ·5 = 37·8
− (calf skinfold)

Height	55.0	56.5	58.0	59.5	61.0	62.5	64.0	65.5	67.0	68.5	70.0	71.5	73.0	74.5	76.0	77.5	79.0	80.5	82.0	83.5	85.0	86.5	88.0	89.5
Humerus	5.19	5.34	5.49	5.64	5.78	5.93	6.07	6.22	6.37	6.51	6.65	6.80	6.95	7.09	7.24	**7.38**	7.53	7.67	7.82	7.97	8.11	8.25	8.40	8.55
Femur	7.41	7.62	7.83	8.04	8.24	8.45	8.66	8.87	9.08	9.28	9.49	9.70	9.91	**10.12**	10.33	10.53	10.74	10.95	11.16	11.37	11.58	11.79	12.00	12.21
Biceps	23.7	24.4	25.0	25.7	26.3	27.0	27.7	28.3	29.0	29.7	30.3	31.0	31.6	**32.2**	33.0	33.6	34.3	35.0	35.6	36.3	37.1	37.8	38.5	39.3
Calf	27.7	28.5	29.3	30.1	30.8	31.6	32.4	33.2	33.9	34.7	35.5	36.3	37.1	**37.8**	38.6	39.4	40.2	41.0	41.8	42.6	43.4	44.2	45.0	45.8
SECOND COMPONENT	½	1	1½	2	2½	3	3½	4	**4½**	5	5½	6	6½	7	7½	8	8½	9						

Weight (lb.) = 175·7
Ht. $\sqrt[3]{\text{Wt.}}$ = 13·26

Upper limit	11.99	12.32	12.53	12.74	12.95	13.15	13.36	13.56	13.77	13.98	14.19	14.39	14.59	14.80	15.01	15.22	15.42	15.63
Mid-point	and	12.16	12.43	12.64	12.85	13.05	**13.26**	13.46	13.67	13.88	14.01	14.29	14.50	14.70	14.91	15.12	15.33	15.53
Lower limit	below	12.00	12.33	12.54	12.75	12.96	13.16	13.37	13.57	13.78	13.99	14.20	14.40	14.60	14.81	15.02	15.23	15.43
THIRD COMPONENT	½	1	1½	2	2½	3	**3½**	4	4½	5	5½	6	6½	7	7½	8	8½	9

	FIRST COMPONENT	SECOND COMPONENT	THIRD COMPONENT	
Anthropometric Somatotype	2	4½	3½	BY: E.M.
Anthropometric plus Photoscopic Somatotype				RATER:

Figure 2.1 Heath–Carter somatotype rating form with mean values for basketball players. (The form depicted is reprinted from Heath and Carter, 1967 by permission of The Wistar Institute Press. The data recorded on the form are from the 1968 Mexico Olympics.)

As a further aid in this study, a standard somatotype photograph (Sheldon *et al.*, 1954) was taken on 5 × 7-inch (12.7 × 17.8 centimeter) film using a camera with a subject–lens distance of 14 feet 9 inches (4.5 meters). The background grid and vertical markers were as described by Dupertuis and Tanner (1950). The subjects were nude except for a small triangle of cloth covering the pubic region. Contact prints were made from the negatives. The photographs were used for somatotype ratings, illustration, and race classification.

Somatocharts are used in this study to graphically display somatotype data and as a basis for determining the patterns of component dominance. The somatochart first devised by Sheldon (1940) is a schematic, triangular shaped, two-dimensional representation of the theoretical range of known somatotypes. It shows the distribution of somatotypes plotted as points and their concentrations on the somatochart. However, when using open-ended rating scales, as in the Heath–Carter method of somatotyping, extrapolation beyond the original triangle is used. A somatochart which allows for extrapolation to new somatotypes is used for presenting data in this chapter and conforms with the principles used by Roberts and Bainbridge (1963).

The somatochart is divided into sectors by three axes which intersect at the center of the "triangle." These sectors and the somatotypes in them are given names according to the relative rank or dominance of the components of the somatotype. For example, the 3–5–2 somatotype

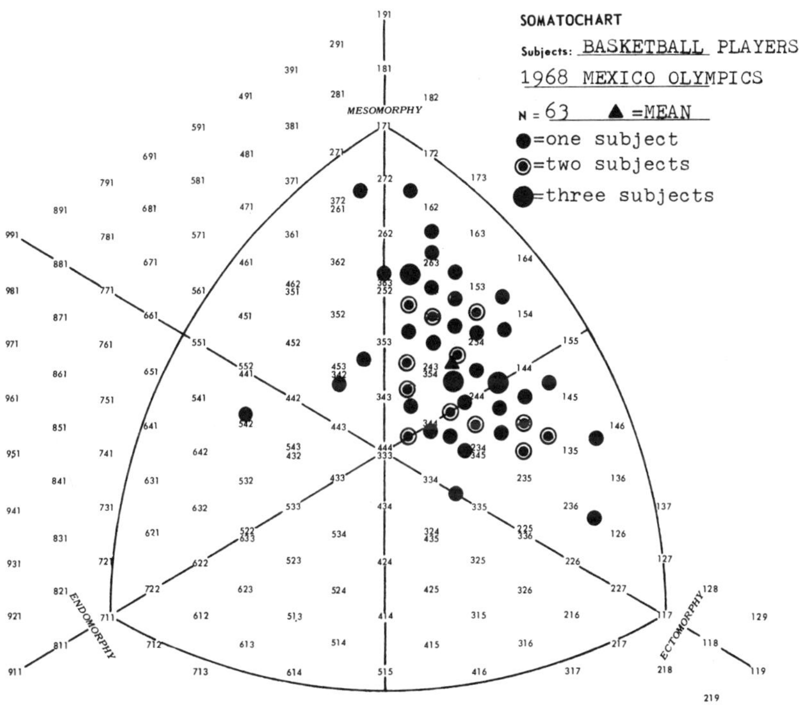

Figure 2.2 A somatochart showing the somatotype distribution of Olympic basketball players. (The form depicted is reprinted from Carter and Phillips, 1969 by permission of the *Journal of Applied Physiology*. The data recorded on the form are from the 1968 Mexico Olympics.)

is called an endo–mesomorph because mesomorphy is dominant, with endomorphy second in dominance. A 1–6–3 is called an ecto–mesomorph, a 2–3–5 a meso–ectomorph, a 2–4–4 an ectomorph–mesomorph, and a 2–5–2 a balanced mesomorph.

The somatochart (Figure 2.2) is a useful framework for comparing the frequencies of somatotypes falling in or on the named parts of the chart. A diagrammatic representation of the usual categories named for comparative analysis is presented in Figure 2.3. These usual categories were modified for use in this study to provide for fewer groups and to take into consideration the distributions of the male and female athletes. The groups are shown in Figure 2.4 and are defined as follows:

1. *Central:* No basic dominance. Ratings are combinations of 3, 4, or 5.

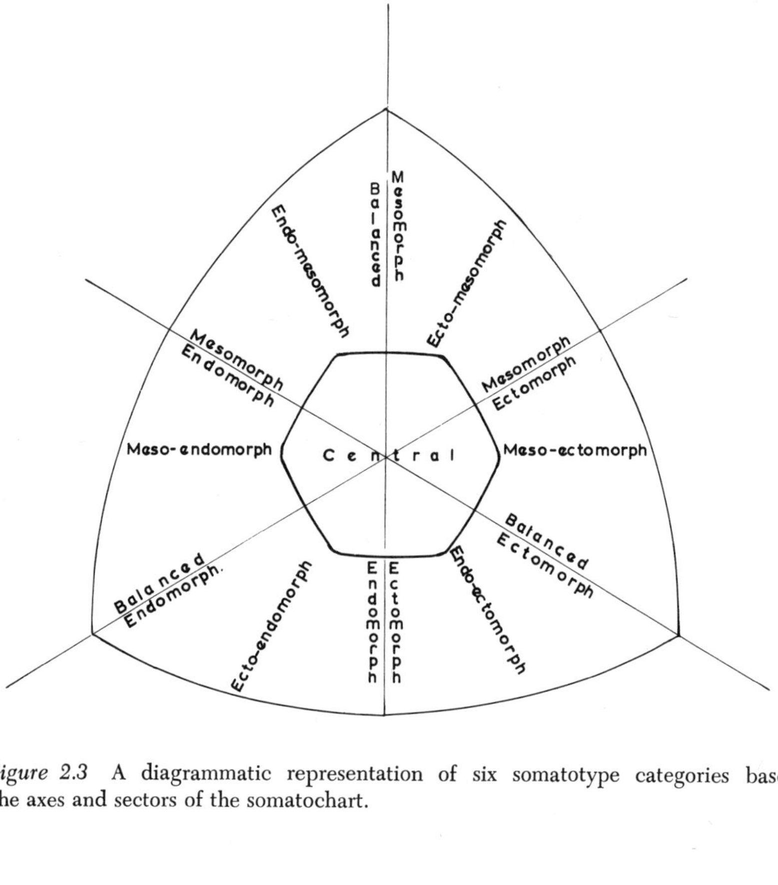

Figure 2.3 A diagrammatic representation of six somatotype categories based on the axes and sectors of the somatochart.

2. *Balanced mesomorphy:* Mesomorphy is two units greater than the other components. The mesomorphic rating must be 5 or greater and the first and third component ratings may not differ by more than 1.
3. *Endomorphy–mesomorphy:* First and second components are dominant. The rating of one of them is at least 4 with the third component 2 units or more lower.
4. *Ectomorphy–mesomorphy:* Second and third components are dominant. One of them is at least 4, with the first component 2 units or more lower.
5. *Endomorphy:* Endomorphy dominates and it is at least 2 units greater than the second component.
6. *Ectomorphy:* Ectomorphy dominates and it is at least 2 units greater than the second component.

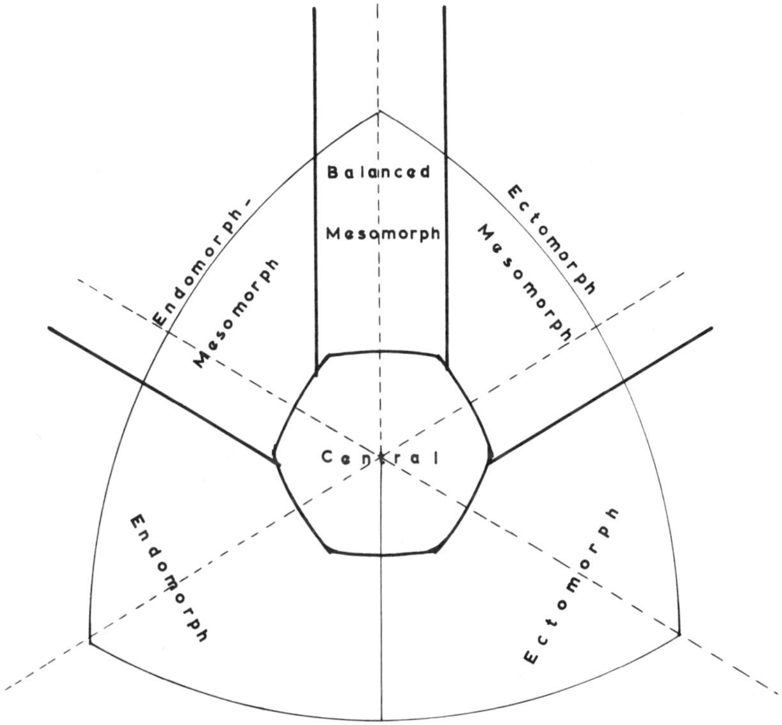

Figure 2.4 A diagrammatic representation of six somatotype categories based on major component dominance.

2. *Somatometry*

In order to characterize the structure and size of athletes and to study the relationships between body regions, a number of anthropometric measurements were taken directly on the individual. The athletes were nude except for a minimal covering of the genitals.

The measurements used in this study were the internationally accepted techniques of Martin and Saller (1957) and Tanner (1964).

The instruments used and the measurements taken are described in Appendix B. Although a large number of measures were taken, it was not possible to analyze all of them in the present report. Selected measurements are analyzed here and others will be utilized in later studies.

B. *Classification of the Athletes*

Although 1265 athletes were measured in this study, complete anthropometric data were not taken on all. Forty-five athletes (3.6%) are deficient in data on some of the variables analyzed. The number of athletes according to sport, sex, the percentage in our sample of the total number, and the number of medalists by sport are presented in Table 2.1.

TABLE 2.1. NUMBER OF ATHLETES BY SPORT, SEX, AND MEDALS

SPORT	SEX	Total number of Athletes by sport specialities	Number of Athletes studied	Percentage of Athletes studied	Number of medalists
Track and Field	M F	864 258	317 82	36.69 31.78	41
Gymnastics	M F	139 115	28 21	20.14 18.26	6
Weightlifting	M	174	58	33.33	7
Rowing	M	394	87	22.08	18
Mod. Pentathlon	M	61	24	39.34	1
Basketball	M	191	63	32.98	4
Water Polo	M	163	71	43.56	1
Cycling	M	361	105	29.08	3
Wrestling	M	320	90	28.12	11
Boxing	M	337	141	41.84	16
Canoeing	M F	171 37	49 6	28.65 16.22	6
Swimming and Diving	M F	331 252	84 39	25.38 15.48	6
T O T A L		4168	1265	30.35%	120

Thirteen sports (with diving considered separately from swimming) are represented, with women in five of them. The women's sample of 148 is 11.7% of our total sample and 22.4% of the women in the five sports. Our sample of 1117 men is 31.4% of the men in 13 sports, indicating that we did slightly better at sampling the men than at sampling the women.

The athletes were grouped by four major categories for the purposes of analysis: sex, sport, event, and racial group. The first two categories, sex and sport, presented no difficulties because the athletes were self-defined in the specific groups. The latter two categories, event and race, provide less assurance of valid determination of category.

The 1265 athletes were spread over 129 separate Olympic events (Greco–Roman and Freestyle Wrestling counted separately), with the numbers too low in some events to provide any meaningful analysis. For some sports, however, such as basketball, waterpolo, and modern pentathlon, no breakdown of the group was necessary. For other sports, the many Olympic events were regrouped into relatively homogeneous larger groups in order to facilitate analysis. The groupings provided our working hypothesis for testing the relationships between structure and function. (There will be disagreement with any grouping, but *one* has to be chosen, keeping in mind the number of subjects, the number of variables, and the depth and kind of the proposed analysis.)

In general, the following bases for regrouping were used: (*1*) Events with low numbers of subjects but similar function were combined; e.g., light–heavyweight ($N = 4$) and heavyweight ($N = 3$) wrestlers were combined as "light–heavy and heavy" ($N = 7$). (*2*) Events with essentially the same function, regardless of number of subjects, were combined; e.g., individual road race, cycling ($N = 38$), and road team time trial, cycling ($N = 31$) were combined as "road" ($N = 69$). (*3*) Previous studies in some sports indicated that some events were similar to each other, while other events were dissimilar; e.g., 800-meter and 1500-meter track for men are similar to each other; and weight throwing (shot, discus, hammer) is different from javelin throwing. The athlete himself designed his primary event at the Games, if he was entered in more than one event. Usually no conflict arose when the athlete subsequently did better in his secondary event, as this was likely to be in the same grouping as the primary event.

Categorization of populations into racial groups depends on the system and criteria used. For this study, the racial groupings were determined by Drs. Johanna Faulhaber and Venerando Correnti following the criteria of Comas (1960). The classifications were based on photographs (both somatotype and identification photographs) and besides skin color the physical characteristics rated were shape of the body in general; proportions of the segments of the limbs; facial structure; form of eyes, lips, and nose; color and texture of hair. The results of this classification are shown in Table 1.2. For this study the athletes were grouped into four major categories for analysis: Caucasoid, Mongoloid, Mestizo (Indian plus Caucasoid), and Negroid (the latter group including groups a,b,c,d in Table 1.2). There were some athletes whose racial background was unknown, and these were withheld from racial analysis.

Data were also obtained on reference groups of Mexican nonathletes, 267 males and 96 females. Descriptive data from these samples are included but no Statistical comparisons with the athletes are made (see also pp. 49, 67-69).

C. *Choice of Variables*

Of the variables measured, only a limited and representative number were used for the descriptive and comparative analysis. The 13 measurements used for descriptive analysis were taken to be representative of the major dimensions of the body in terms of size and shape, somatotype determination, and their usefulness in studying indices. A list of the variables is given in Table 2.2. The reference numbers refer to the definition of the variable given in Appendix B in Chapter 3. Comparative analysis was performed on 10 of the 13 variables in each event-group and sport.

TABLE 2.2. LIST OF VARIABLES ANALYZED

Number of variable	Name of variable	Reference number in Appendix B
1.	Age in years	
2.	Weight	53
3.	Stature, maximum stretch	1
4.	Trunk length	36
5.	Biacromial breadth	20
6.	Biiliocristal breadth	24
7.	Upper+lower arm length, right	64
8.	Total leg length, right	72
9.	Sum of three skinfolds (Σ of triceps, subscapular and suprailiac skinfolds)	
10.	Height-weight ratio $\quad HWR = \dfrac{\text{variable (3) in inches}}{\sqrt[3]{\text{variable(2) in pounds}}}$	
11.	Endomorphy) From the Heath-Carter somatotype rating form	
12.	Mesomorphy) From the Heath-Carter somatotype rating form	
13.	Ectomorphy) From the Heath-Carter somatotype rating form	

The 10 variables selected for analysis were (in addition to age), the two main indicators of body size (height and weight), the three somatotype components (endomorphy, mesomorphy, and ectomorphy) as indicators of body form and composition, two breadth indicators (biacromial and biiliocristal), and three length indicators (arm, leg, and trunk length). The analyses for endomorphy and ectomorphy were made using the sum of three skinfolds (triceps, subscapular, and suprailiac), and the height–weight ratio, respectively. This procedure was used for a number of reasons: (*1*) Endomorphy is derived directly from the skinfolds; ectomorphy is derived directly from the HWR; (*2*) both the skinfolds and the height–weight ratios are more precise than the somatotype scales, which have discrete half-unit intervals; and (*3*) use of the original measurement scales for skinfolds and the HWR reduces skewness occurring in some samples. For example, weight lifters have very low HWR's and their somatotype ratings for the third component (ectomorphy) tend to "pile up" on ratings of $\frac{1}{2}$ and 1. Their mean is 1.0, the standard deviation is 0.48, and the range is from 0.5 to 2.0 units of ectomorphy, whereas the HWR itself is more normally distributed with a mean of 12.06, a standard deviation of 0.42, and a range from 10.85 to 12.73. A similar effect is seen in some groups with low skinfolds. On the other hand, the same effect is not apparent in mesomorphy because of the open-ended scale at the upper end. Furthermore, one cannot use the same procedures with mesomorphy because it is a composite of height, bone breadths, limb circumferences, and skinfold corrections.

Although not a structural variable, age was analyzed because of its usefulness as a biological reference point. Only descriptive analysis was carried out on age.

D. *Statistical Procedures*

The majority of data were analyzed on a Control Data Corporation 1604 computer using the Biomedical Computer Programs of the University of California (Dixon, 1968). The descriptive data were computed using program 01D. The standard deviation calculated by this program is an estimate of the population value sigma (σ). To obtain the sample standard deviation(s), the values for sigma given in the tables should be multiplied by $(N-1)/N$, where N is the sample size. The analysis of variance and comparisons among more than two means were made using the program 07V. The range values for Duncan's (1955) New Multiple Range Test were used in the tests of significance. The 5% level of confidence was used to test the significance of the differences between the means, although both the 1% and 5% levels are given in most tables. The F-test is regarded as a fairly robust test and can tolerate rather large differences in sample and variance sizes. The sample variances were examined for equality before applying the *F*-test and where a particular variance was different from the remainder, that sample was withheld from the analysis in question. On some occasions, the analysis was run both with and without the sample. For comparisons between two means, a one-way analysis of variance, program 01V, was used. Chi-square tests were calculated to test the differences between some somatotype frequencies on the somatocharts.

The descriptive analysis made use of the total number of athletes with data for the variable being analyzed. The analysis of variance programs were run using the number of subjects in each event group who provided complete data on all 10 variables.

Bibliography

Azuma, T. (Ed.) (1964). "Olympic Medical Archives, Report, Tokyo 1964." The Japanese Olympic Medical Archives Committee, Tokyo, Japan.

Carter, J. E. L., and Phillips, W. D. (1969). Structural changes in exercising middle-aged males during a 2-year period. *J. Applied Phys.* **27**, 787–794.

Comas, J. (1960). "Manual of Physical Anthropology." Charles C Thomas, Springfield, Illinois.

Correnti, V., and Zauli, B. (1964). "Olimpionici 1960." Marves, Roma, Italy.

Cureton, T. K., Jr. (1951). "Physical Fitness of Champion Athletes." Univ. of Illinois Press, Urbana, Illinois.

Dixon, W. J. (Ed.) (1968). "Biomedical Computer Programs." Univ. of California Press, Berkeley, California.

Duncan, D. B. (1955). Multiple range and multiple *F* tests. *Biometrics* **11**, 1–45.

Dupertuis, C. W., and Tanner, J. M. (1950). The pose of the subject for photogrammetric anthropometry, with special reference to somatotyping. *Am. J. Phys. Anthrop.* **8**, 27–47.

Gedda, L. (1961). Chapter 5, Sports and genetics. A Study on twins (351 Pairs). *In* "Health and Fitness in the Modern World" (L. Larson, ed.). The Athletic Institute, Chicago, Illinois.

Heath, B. H., and Carter, J. E. L. (1967). A modified somatotype method. *Am. J. Phys. Anthrop.* **27**, 57–74.

Hirata, K. (1966). Physique and age of Tokyo olympic champions. *J. Sports Med. and Phys. Fitness* **6**, 207–222.

Jokl, E. (1964). Essay on medical sociology of sports. *In* "Medical Sociology and Cultural Anthropology of Sport and Physical Education." Charles C Thomas, Springfield, Illinois.

Keys, A., Brozek, J., Henschel, A., Mickelson, O., and Taylor, H. L. (1950). "The Biology of Human Starvation." Univ. of Minnesota Press, Minneapolis, Minnesota.

Kohlraush, W. (1929). Zuzammenhange von Korperform und Leistung. Ergebnisse der anthropometrischen Messungen an der Athleten der Amsterdamer Olympiade. *Arbeitphysiol.* **2**, 187–204.

Lasker, G. W. (1947). The effects of partial starvation on somatotype. *Am. J. Phys. Anthrop.* **5**, 323–341.

Martin, R., and Saller, K. (1957). *Lehrbuch der Anthropologie.* I. Fischer, Stuttgart.

Parnell, R. W. (1958). "Behaviour and Physique." Edward Arnold Ltd., London.

Roberts, D. F., and Bainbridge, D. R. (1963). Nilotic physique. *Am. J. Phys. Anthrop.* **21**, 341–370.

Sheldon, W. H. (1965). Need for a somatotype taxonomy. *Maudsley Bequest Lecture.* Royal Society of Medicine, London.

Sheldon, W. H., Dupertuis, C. W., and McDermott, E. (1954). "Atlas of Men." Harper Brothers, New York.

Sheldon, W. H., Stevens, S. S., and Tucker, W. B. (1940). "The Varieties of Human Physique." Harper Brothers, New York.

Tanner, J. M. (1964). "The Physique of the Olympic Athlete." George Allen and Unwin, London, England.

Tittel, K. (1965). Zur Biotypologie und funktionellen Anatomic des Leistungssportlers. *Nova Acta Leopoldina,* **30**, 172.

Worringer, K. A. (1927). *Was muss der Arzt von den Leibesunbungen wissen,* Munchen, Germany. (Cited in Correnti and Zauli, 1964.)

3

Results of Physical Anthropology Analysis

I. Analysis within Sports

The results of the analysis are presented by sport, using descriptive data, frequencies of somatotypes, differences within sports, and summaries of somatocharts. The order of presentation is as follows:

1. Track and Field (Men)
   a. Track
   b. Field
2. Track and Field (Women)
3. Swimming (Men)
4. Swimming (Women)
5. Diving (Men)
6. Diving (Women)
7. Water Polo
8. Modern Pentathlon
9. Basketball
10. Rowing
11. Canoeing (Men)
12. Canoeing (Women)
13. Cycling
14. Gymnastics (Men)
15. Gymnastics (Women)
16. Wrestling
17. Boxing
18. Weight Lifting
19. Reference (Men)
20. Reference (Women)

For each event grouping, there is (*1*) a descriptive data table showing the number of subjects, mean, standard deviation, and minimum and maximum scores for each of 13 variables; (*2*) a frequency table of the somatotypes according to component dominance based on the somatochart groupings described on pp. 18–19; (*3*) an analysis of variance table of the differences between event-groups within the sport on 10 variables taken one at a time (where there is more than one group); and (*4*) photographs of selected athletes from various events. In the analysis of variance tables, the size of the *F*-ratio tells us whether there is or is not a significant difference between the group means for the variable under analysis. On the righthand side of the table, the group means for each variable are ordered from left to right, from lowest to highest values. The groups underlined by a common line do not differ significantly. Those not underlined by a common line do differ significantly. If groups are not underlined, each group differs from every other group. Occasionally, there is an apparent discrepancy between a nonsignificant *F*-ratio and a significant group difference. Probably this is due to the different power or chance factors operating within the *F*-test and Duncan's Multiple Range Test. Such discrepancies are discussed where they occur.

A. *Track and Field* (*Men*)

1. *Track*

This group includes the runners, hurdlers, and walkers (see Tables 3.1–3.9 at the end of this chapter). It is the largest of our athletic samples, with $N = 246$. The 110-meter hurdlers are included with the sprinters because their event primarily demands sprinting speed, to which is added the hurdling skill. Similarly, the 400-meter hurdlers primarily have to be good 400-meter runners who can also handle the hurdling technique.

Age. There seems to be little relationship between age and event. Individually, the ages range from 16 to 42 years, with the youngest athletes in sprinting, 400 meters, and middle distance; but the oldest athletes are also in sprinting, as well as the distance runs and walking.

Somatotype. The sprinters are significantly greater on endomorphy and mesomorphy than all other groups except walkers, and are significantly lower on ectomorphy than all other groups. The long distance runners are significantly less mesomorphic than the walkers, while all other differences are not significant (see Table 3.9; Figures 3.1 and 3.2). The skinfold values show that all the track groups are low, but

Figure 3.1 Marathon runner (Number 329), Negroid, Ethiopia, age 35, somatotype 1–4–4½; mean for Marathon runners, 1.4–4.3–3.5.

Figure 3.2 Walker (Number 4) Caucasoid, U.S.A., age 37, somatotype 1½–4½–3½; mean for walkers, 1.6–4.7–3.4.

there are more sprinters with skinfolds greater than 25.0 millimeters than other track athletes. Some of the leanest athletes in our sample are track athletes, but most are in the longer distances. The lowest skinfold recording is 11.2 millimeters for a distance runner, but one sprinter is the next lowest, with 11.7 millimeters. The most mesomorphic men are in the sprinting group, with ratings as high as 8.0 for a few athletes. In other track groups, the highest value is 6.0. We may summarize the average track somatotype as $1\frac{1}{2}$–$4\frac{1}{2}$–$3\frac{1}{2}$.

Generally speaking, all track groups have similar somatotype distributions and are concentrated mainly in the ecto–meso category. All athletes are relatively far from the endomorphic corner, but show a considerable range on the scales of mesomorphy and ectomorphy. Table 3.8 shows that the long-distance track runners and walkers (1500 meters and greater) have fewer mesomorphs and more ecto–mesomorphs than the short-distance runners (100–800 meters). Of the eight sprint medalists, only Pender ($1\frac{1}{2}$–7–1) and Carlos ($1\frac{1}{2}$–6–3) are at the edge of the distributions. All other medalists fall reasonably close to the center. Similarly, in other events Wedderburn (400 meters), Farrell (800 meters), Tummler (1500 meters), and Pedraza (20-kilometer walk) are at the edge of their distributions, while other medalists are well within their distributions. Two of the medalists overlap in terms of event. Keino won the gold in the 1500 and was second in the 5000-meter event. Wolde was first in the marathon and second in the 10,000 meter. Each of these individuals is classified in the event in which he won the gold medal. It is interesting to note that two marathon winners from the same country, Bikila and Wolde from Ethiopia, have identical somatotypes at 1–4–$4\frac{1}{2}$.

Body size. The 400-meter and the middle-distance runners are the tallest trackmen, while the marathon and long-distance runners are the shortest. The heaviest trackmen are the sprinters and 400-meter runners and the lightest are the marathon and long-distance runners.

The marathon and long-distance runners have smaller biacromial (shoulder) breadths than the middle-distance runners, 400-meter runners, and sprinters. The walkers do not differ from any of the other groups. For the biiliocristal (hip) breadth, the sprinters have the narrowest breadth and they differ significantly from the walkers, who have the widest breadth. The marathon runners have the shortest arm length; the long-distance runners and walkers have only a slightly longer arm length. The longest arms are those of the sprinters, middle-distance runners, and 400-meter runners, who do not differ from each other. In terms of leg length, there are two groups: The longest legs belong to the sprinters, the middle-distance runners, and 400-meter runners, and the shortest belong to the marathon and long-distance runners and walkers. The walkers have a longer trunk length than the marathon runners, long-distance runners, and the sprinters, while the sprinters, middle-distance runners, and 400-meter runners do not differ from each other. Overall, the largest track athletes are the 400-meter runners, the sprinters, and middle-distance runners, while the smallest are the marathon and long-distance runners. These latter two do not differ significantly on 9 of 10 variables. The walkers tend to be in between these groups except for hip breadth and trunk length, for which they show the greatest values.

2. *Field*

This group contains the jumpers, throwers, pole vaulters, and decathlon men (see Tables 3.8, 3.10–3.16; Figures 3.3–3.6). The decathlon

Figure 3.3 Shot putter, Caucasoid, Poland, age 32, somatotype $2\frac{1}{2}$–$7\frac{1}{2}$–1; mean for weight throwers, $3\frac{1}{2}$–7.1–1.

Figure 3.4 Javelin thrower, Caucasoid, USSR, age 28, somatotype 2–6½–1½; mean for javelin throwers 2–5.6–2.3.

are included in this group because 6 of the 10 events are field events. Although the numbers are relatively small, on the basis of the data in Tanner (1964) and in Correnti and Zauli (1964), the weight throwers (shot-put, discus, and hammer) are separated from the javelin throwers. The high jumpers, long jumpers, and triple jumpers were combined into one group because functionally they all require explosive leg power.

Age. The average ages for the field groups are 24 to 28 years. The individual range is from 17 years (a jumper) to 43 years, with those over 30 years of age being confined to the weight throwers and javelin throwers.

Somatotype. The weight throwers are clearly more endomorphic and mesomorphic, but less ectomorphic than the other four groups. The jumpers, pole vaulters, javelin, and decathlon men have low-average skinfolds, similar to those of the track men, but the weight throwers have a higher mean and a wide range. The range in endomorphy is therefore 1.5 to 6.0 units. In mesomorphy the values are high with a range between 5.0 and 9.0, and in ectomorphy they are uniformly low,

Figure 3.5 Pole vaulter, Caucasoid, United States, age 22, somatotype $1\frac{1}{2}$–$5\frac{1}{2}$–$2\frac{1}{2}$; mean for pole vaulters, $1\frac{1}{2}$–4.8–3.2.

Figure 3.6 High jumper, Caucasoid, USSR, age 21, somatotype $1\frac{1}{2}$–$4\frac{1}{2}$–$4\frac{1}{2}$; mean for high jumpers, 1.7–4.4–3.4.

ranging from 0.5 to 2.5 units. The vaulters, jumpers, javelin throwers, and decathlon men are not different from each other in endomorphy, but the jumpers are less mesomorphic and more ectomorphic than the javelin and decathlon men.

From Table 3.8 we see that the weight throwers are quite different from the remaining groups. They are the only group in all of track and field which are dominant endo–mesomorphs (61%); the remainder are dominant mesomorphs. On the other hand, the jumpers, vaulters, and decathlon men have no dominant endo–mesomorphs. The distribution of the decathlon men overlaps slightly with that of the javelin throwers, and with the jumpers and vaulters. Of the six medalists in the field events, only Lusis, the gold medalist in the javelin throw, and Seagren, the gold medalist in the pole vault, are at the edge of their somatotype distributions—both athletes are the most mesomorphic in their event groups.

Body size. The weight throwers are considerably heavier than the other four groups. For weight (and for skinfolds) the analysis of variance was performed both with and without the weight throwers because of the excessively high variance. When the analysis was performed without weight throwers, the observation—that vaulters, jumpers, javelin men, and decathlon men do not differ—was the same as when weight throwers were included. The weight throwers are significantly larger in shoulder breadth and trunk length than the other four groups. In arm length, the groups do not differ significantly from each other. On hip breadth, the jumpers are narrower than the javelin throwers or weight throwers, and on leg length the jumpers have longer legs than the javelin throwers. For leg length the F-ratio is insignificant, so the mean difference between the jumpers and javelin throwers of 5.2 centimeters should be viewed with caution.

It is interesting to note that the decathlon men who are required to run, jump, and throw consistently rank in the second, third, or fourth places on all 10 variables in the field event analysis. In terms of size, however, they are much taller and heavier than the track athletes.

B. *Track and Field* (*Women*)

All of the major groupings for women's track and field are included in this section (see Tables 3.17–3.24; Figure 3.7). In the analysis, the weight throwers (shotput and discus) are separated from the javelin throwers partly for the same reason as for the male athletes, but also on the basis of a preliminary inspection of the data which indicated that this procedure might be profitable. It should be noted that the category middle distance for women includes 400 and 800 meters and

Figure 3.7 Middle-distance runners.

therefore differs from the groupıng for men. These were the two longest distances run in the 1968 Olympics by women.

Age. The mean age of the various groups ranges from 20 in the middle distance, weight throwers, and javelin throwers, to 23 in the pentathlon athletes. The individual range is from 15 to 29 years, a range common to both sprinters and pentathlon women. In addition, middle-distance runners, jumpers, and weight throwers have 15-year-old athletes in their group.

Somatotype. The weight throwers are significantly more endomorphic, mesomorphic, and less ectomorphic than the other five groups. In addition, the javelin throwers are more endomorphic than the middle-distance runners and jumpers, and the sprinters more mesomorphic and less ectomorphic than the middle-distance runners and jumpers. Table 3.25 shows that the field throwers are dominantly endomorphic or endo–mesomorphic somatotypes; however, the javelin throwers range all the way from a 6–5–1 through two central somatotypes to a 2–2–5. The wide range of endomorphy and ectomorphy, the small number of subjects, as well as the fact that the 6–5–1 (Penes) was the gold medalist at Tokyo and second at Mexico, indicates that one should view with caution the mean values for javelin throwers. Both the sprinters and the middle-distance runners have the same percentage of central somatotypes, but there is an increase in the more ectomorphic somatotypes in the distance runners. However, of the 14 medalists, Ferrell, the most ectomorphic of the sprinters ($1\frac{1}{2}$–$2\frac{1}{2}$–$4\frac{1}{2}$), and Manning, the most ectomorphic of the middle-distance runners at ($1\frac{1}{2}$–2–$5\frac{1}{2}$) as well as the discus thrower (Penes) mentioned earlier, are all at the extremes of their distribution.

Body size. The shortest women athletes are the sprinters, who differ from the weight throwers and javelin throwers. The tallest are the javelin throwers, who differ significantly from all other groups. The analysis for weight was conducted without the weight or javelin throwers. The athletes in both of these groups have high mean weights which made them more than 13 kilograms heavier than any of the other groups. Of the remaining athletes, the middle-distance runners are lighter than the pentathlon athletes. The weight and javelin throwers have broader shoulders than the middle-distance runners and sprinters, and wider hips than the jumpers, sprinters, middle-distance runners, or pentathlon women. The pentathlon athletes have wider hips than the jumpers and sprinters. The longest arms belong to the javelin throwers, who are significantly different from the sprinters and middle-distance runners. The same is true for leg length and for trunk length, but in addition, the javelin throwers have a longer trunk length than the jumpers. The jumpers and sprinters have shorter trunk lengths than the weight throwers and pentathlon athletes. The pentathlon women, who run, jump, and throw, are ranked between the second and fourth places on each of the 10 variables, suggesting that their physique must be generally mid-range in women's track and field for successful performance.

C. *Swimming* (*Men*)

Because the 68 swimmers were spread thinly over 14 different swimming events by distance and stroke, it was decided to analyze the swimmers grouped according to stroke only (see Tables 3.26–3.33; Figure 3.8). For the men, this provided five groups—freestyle, breaststroke, backstroke, butterfly, and medley.

Age. The average age of the male swimmers is 19 years with a mean of 18 years for backstroke and 20 years for both freestyle and butterfly. The individual age range is from 14 years (backstroke) to 25 years (free style).

Somatotype. The backstrokers are significantly less mesomorphic than the medley and breaststroke swimmers, and they are more ectomorphic

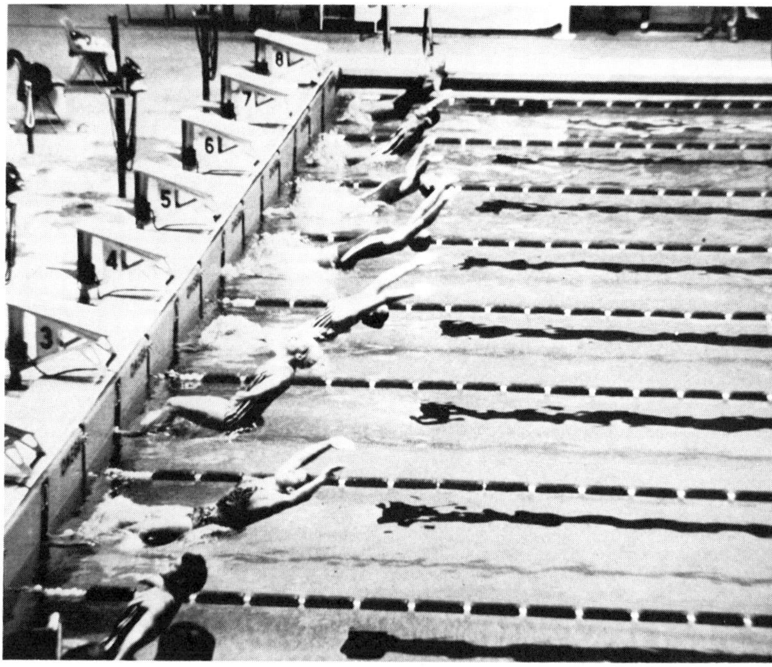

Figure 3.8 Swimmers.

than the medley, breaststroke, or butterfly swimmers. The mean somatotype of the backstrokers (2–4½–3½) is a little different from the other four groups, whose basic pattern is 2–5–3. More than half the swimmers (53%) are dominant mesomorphs, with ecto–mesomorphy the next highest group, and central third in rank. Neither of the two medalists (Munoz—200-meter breaststroke, or Ferris—200-meter butterfly and 200-meter individual medley) have somatotypes which are atypical for their groups.

Body size. There are no significant differences between the groups in weight, biacromial breadth, arm length, or trunk length. The freestylers are taller, have wider hips and greater leg length than the breaststrokers. The backstrokers also differ from the breaststrokers on leg length. It should be noted, however, that the differences on ectomorphy (discussed earlier) and leg length are only of borderline significance considering the nonsignificant F-ratio.

D. *Swimming (Women)*

As with the men, the small number of women swimmers forced us to use stroke rather than stroke and distance for our group analysis (see Tables 3.34–3.40). We arrived at four groups—freestyle, breaststroke, backstroke, and medley. There are no butterfly swimmers in our sample.

Age. The average age of the female swimmers is 16 years of age, with an individual range from 12 to 23 years. The groups average 15 to 18 years, with the youngest swimmers in freestyle and breaststroke, and the oldest in breaststroke.

Somatotype. Apart from one very endomorphic breaststroke competitor, the swimmers are largely central somatotypes, or endomorph–mesomorphs. The breaststroke, backstroke, and medley swimmers averaged 3½–4–3 in somatotype, while the freestyle swimmers averaged 3–3½–4. The higher ectomorphy of the freestyle swimmers is significantly different from the means of the other three groups. To equalize the variances on skinfolds, the very endomorphic subject was removed from the breaststroke subgroup. The F-ratio is not significant, but the multiple range test indicates that the medley swimmers may be slightly more endomorphic than the breaststroke swimmers. The gold and silver breaststroke medalist, Bjedor, a 3½–3½–3 somatotype, is more towards the center of the somatochart than any of the other swimmers in her group, but is relatively central when all swimmers are considered.

Body size. As can be seen from Table 3.40, there are no differences between any of the groups in terms of body size variables. There is a slight tendency only in terms of ranking for the medley swimmers to be the largest of the swimmers.

E. *Diving (Men)*

The 16 divers include both springboard and tower divers, as most divers compete in both events at the Olympics (see Tables 3.41 and 3.43). As shown in Table 3.43, the divers range over only two somatotype groups—69% in dominant mesomorphy, and 31% ecto–mesomorphy. They tend to be relatively lean, averaging 20.0 millimeters in skinfolds. The gold and silver medalist, Dibiasi, whose somatotype is 2–5–2½, is at the center of the somatotype distribution of the divers.

F. *Diving (Women)*

The small number of women divers included both tower and springboard divers (see Tables 3.39 and 3.42, and Figure 3.9). All of the divers except one fall in the central somatotype group, with all but one of the physiques being higher than the center of the somatochart.

Figure 3.9 Diver.

G. *Water Polo*

It appears that water-polo players range from small to large and from lean to fat (see Tables 3.43 and 3.44). Table 3.43 shows that the majority (75%) of water-polo players are either endo–mesomorphic or dominant mesomorphs with smaller but equal numbers of ecto–mesomorphs and central somatotypes. Some of the endo–mesomorphs are extreme with very high ratings in both endomorphy and mesomorphy. For two of the athletes, endomorphy dominates over mesomorphy.

H. *Modern Pentathlon*

The silver medalist in the team event, Shaparris, has a somatotype of $2\frac{1}{2}$–5–2, and although on the endomorphic side of the mesomorphic axis, is not far from the center of the distribution. The somatochart distribution shows a fairly broad distribution over the whole mesomorphic sector, with the concentration more towards the ectomorphic side (see Tables 3.43 and 3.45).

I. *Basketball*

The wide spread in height and in weight shows that although many players in basketball are extremely large, relatively small players still have their place (see Tables 3.43 and 3.46). All physiques but one are dominant in either mesomorphy or ectomorphy. Forty percent of the somatotypes are ectomorph–mesomorphs, with central physiques being the next largest group (32%). Almost all of the central physiques are on the ecto–mesomorphic, or meso–ectomorphic side of the somatochart. The four medalists in basketball are from the second place Yugoslavian team. One of these players, Cosic, is the most linear of all of the basketball players, having a somatotype of 1½–2½–6. The other three players fall within the general boundaries of the distribution.

J. *Rowing*

The rowers were grouped into the four basic styles of rowing for analysis—sculls, pairs, fours, and eights. On a number of occasions it was possible to measure crew members in the same boat (see Tables 3.43 and 3.47–3.52).

Age. The youngest rower is in the fours, and the oldest in the pairs. The oldest rower in the eights is 28 years, while in the other three events there are rowers who were 30 years or older.

Somatotype. There are no significant differences in the three components between any of the four groups of rowers. The average somatotype of all groups is close to a 2–5½–2½. Sixty-five percent of the rowers are dominant in mesomorphy with ecto–mesomorphs being the next dominant group (19%). In the single sculls, Wienese, the gold medal winner, is the most mesomorphic of the group as well as of all of the rowers, with a somatotype of 1½–7–1½. The double sculls silver medalists from the Netherlands, Van Dis and Droog, are the most ectomorphic of this group, and are very similar in somatotype. Two of the remaining three doubles have similar somatotypes, while the third double has dissimilar somatotypes. The somatotypes of the pair-oarsmen are distributed fairly evenly around the mesomorphic axis. The bronze medalist, Christianson, is the most ecto–mesomorphic of the oarsmen, with a 1½–4½–3½ somatotype. In the fours we have measured oarsmen from the gold and bronze medal crews. The somatotypes of the four gold medalists from New Zealand extend from the most mesomorphic to a central ecto–mesomorph. The second place Hungarian crew (fours without Cox) is much more concentrated than the New Zealand crew with a narrower range on somatotype units. The four other medalists from two different crews are within the general area of the distribution for this group. The distribution of the fourth place New Zealand eight-oar crew is across the center of the distribution for the total group.

Body size. The groups in rowing do not differ significantly from each other in height, weight, breadths, or lengths. If there is any trend to be seen from the data, it may be that the eights tend to be slightly larger individuals than those from the other crews. The eights ranked highest on all of the size variables.

K. *Canoeing (Men)*

The male canoeists were divided into kayak 1-, 2-, and 4-man, and Canadian 1- and 2-man groupings (see Tables 3.43, 3.53–3.56). A one way analysis of variance was used to test the difference between these two groups (Table 3.56).

Age. The canoeists average 24 years of age, with a range from 18 to 38 years. The Canadian canoeists appear to be slightly younger, having a mean age of 22 years, range 18 to 31, and the kayak canoeists average 25 years and have a range of 19 to 38 years.

Somatotype. The kayak and Canadian canoeists do not differ in terms of somatotype. All three component means are essentially the same, and the average of all canoeists is 2–5½–2½. The somatochart distribution ranges from extreme mesomorphs to a few physiques dominant in ectomorphy. One kayak participant has a 5–6–1 physique, but is atypical of the group. Near the other extreme is the gold medalist from Tokyo, Peterson, whose somatotype is 1–3½–4. Sixty-one percent of the canoeists are mesomorphs, with ecto-mesomorphs being the group second in dominance (29%).

Body size. The kayak canoeists are significantly taller, heavier, have longer trunks, arms, and legs than the Canadian canoeists. The groups do not differ in shoulder breadth and hip breadth. It appears that although the two groups do not differ in body form and composition (somatotype) the kayakists are larger in terms of gross body size and limb length.

L. *Canoeing (Women)*

Unfortunately, the number of female canoeists available for analysis is extremely small (see Tables 3.39 and 3.57; Figure 3.10). There are two singles competitors and two pairs competitors. All four canoeists are dominant in mesomorphy. One of the most mesomorphic of the women athletes is a 4–6–1½. There are three endo-mesomorphs and one mesomorph in this small group.

M. *Cycling*

The cyclists were divided into three groups for analysis—sprinting, pursuit, and road (see Tables 3.43, 3.58–3.62; Figure 3.11).

Age. The distribution is very similar in each of the groupings.

Somatotype. The three groups of cyclists do not differ significantly

Figure 3.10 Canoeists.

Figure 3.11 Cyclist (No. 311), Caucasoid, Sweden, age 25, somatotype $1\frac{1}{2}$–4–4; mean for cyclists, 1.8–4.9–2.7.

on any of the three somatotype components. Particularly notable is the low range on endomorphy, but there is greater variation in the combinations of mesomorphy and ectomorphy. There are very few endo–mesomorphs and the predominant group are mesomorphs (50%) with ecto-mesomorphs at 38%, the next largest group. Both the bronze medalist in the 1000-meter time trial, Kierzkowski, and the gold medalist from Tokyo, Daler, in the 4000-meters individual pursuit, have somatotypes in the center of their respective distributions. In the road team time trial, we measured one of the Pettersson brothers from Sweden, who was one of the two most ectomorphic of our sample with a somatotype of 1½–3½–4.

Body size. There are no significant differences on height, weight, breadths, or lengths between any of the three cycling groups. This analysis shows that the cyclists, when examined in this grouping, are a most homogeneous sports group.

N. *Gymnastics (Men)*

As Olympic gymnastics are essentially all-round, for individual and team events, no attempt to assess characteristics of specialization seemed necessary (see Table 3.63). The group is fairly evenly divided between mesomorphy and ecto–mesomorphy, but the level of endomorphy remains consistently low in all subjects. The gold medalist on the pommelled horse, Cerar, is one of the leanest and most mesomorphic gymnasts, with a somatotype of 1–7–2½.

O. *Gymnastics (Women)*

The majority of the somatotypes (62%) are central physiques, but nearly all of these are in the mesomorphic sector (see Tables 3.39, 3.64, and frontispiece). The five Czechoslovakian bronze medalists (team event) are generally further from the endomorphic sector on the somatochart, and therefore leaner, than most of the other gymnasts. Two of them are also at the highest level (5) of mesomorphy in this group, but one is also at the lowest level (3). The latter gymnast is also the most ectomorphic.

P. *Wrestling*

In the Olympic Games, competition is held in two styles of wrestling, freestyle and Greco–Roman, both with the same weight categories (see Tables 3.65–3.74). The first problem was to see if there were differences between the two styles. A preliminary examination of the somatotype distributions and the descriptive data in Tables 3.66 and 3.67 suggested that there were no significant differences between the two styles. An analysis of variance testing the significance of the difference between the means for both styles on 10 variables is presented in Table 3.68. All F-ratios are low and no significant differences were observed. The differences between the frequencies of the somatotype groupings of the freestyle and Greco–Roman wrestlers were tested by the chi-square technique (see Table 3.65). The obtained Chi-square value of 4.34 was not sufficient to reject the hypothesis of no difference between the distributions. On these bases, further analysis was made on the combined styles to give a total of 90 wrestlers. The wrestlers were then subdivided into pairs of weight classes to provide adequate numbers for analysis (see Table 3.74). In effect, there are two freestyle weights and two Greco–Roman weights included in the single category; for example, freestyle fly and bantam, and Greco–Roman fly and bantam are combined to make a single category of fly–bantamweight wrestlers.

Age. All groups have wrestlers over 30 years of age.

Somatotype. The mean somatotype of all wrestlers is approximately $2\frac{1}{2}$–$6\frac{1}{2}$–$1\frac{1}{2}$. On endomorphy there is an increase of almost two units between the lightest and the heaviest weight classes; on mesomorphy there is an increase of one unit, and on ectomorphy a decrease of one unit. Table 3.74 shows that the two lightest classes are significantly less endomorphic than the two heavier classes, and that the heavy class are more endomorphic than the welter–middle class. In mesomorphy the two lightest classes are significantly lower than the two heaviest classes. The least ectomorphic are the heavyweights, who do not differ from the welter–middleweights, but both are less ectomorphic than the lightest weight group, and the feather–lightweight class is less ectomorphic than the fly–bantamweight, but more ectomorphic than the heavyweight class. Table 3.65 shows that almost two-thirds of the wrestlers are dominantly mesomorphic. Almost one third are endo–mesomorphs. One flyweight, two bantamweights, and one featherweight make up the four ecto–mesomorphic wrestlers. All but four of the extreme endo–mesomorphs occur in the two highest weight classes. The range on endomorphy gets greater from the lowest to highest weight classes. Mesomorphy has a fairly consistent range over all weight classes except a narrow range from 6.5 to 8 units in the heaviest weight class. On ectomorphy the range is largest in the lightweights and smallest again in the heavyweights, where it ranges between one-half unit and 1.5 units. All weight classes have somatotypes with mesomorphic ratings of 7 to 9 units. The freestyle flyweight gold medalist, Nakata, has a somatotype $1\frac{1}{2}$–6–$1\frac{1}{2}$, and Baciu, the silver medalist in Greco–Roman bantamweight, has a somatotype $1\frac{1}{2}$–$6\frac{1}{2}$–$1\frac{1}{2}$. Although not placed in the Mexico Olympics, Imre Polyak from Hungary had previously won four medals at consecutive Olympics in Greco–Roman style and competed in the bantamweight freestyle—his somatotype is $2\frac{1}{2}$–$6\frac{1}{2}$–$1\frac{1}{2}$. These three somatotypes are close to each other and central for their particular weight categories. In the welter and middleweight category, there is a tendency for greater endomorphy to begin. However, the two medalists, Vesper, gold medalist in welterweight Greco–Roman, has a somatotype of 2–7–1½, and Metz, gold medalist in the middleweight Greco–Roman, has a somatotype 2–5½–2. Both these somatotypes are relatively balanced mesomorphs. Four of the five medalists in the feather–lightweight classes are balanced mesomorphs with mesomorphy ranging from 6 to 8½. The gold medalist from Melbourne and Rome, Dagistanli, has a somatotype of 5–7–½. The highest endomorphy of the other four medalists was 2, and this additional 3 units in endomorphy may have been one of the factors against a repeat performance by this 36-year-old wrestler from Turkey. In the small number of heavyweights, endomorphy ranges from 2 to 6 units, but most of the competitors have high endomorphy, as indicated by the mean of 3.5. The bronze medalist, Kmert, has a somatotype of 6–8–½.

Body size. Because the weight categories are by definition discrete and, therefore different, no analysis was carried out on weight. For all of the size variables, height, shoulder and hip width, arm, leg and trunk length, all weight groups differ significantly from each other. The dimensions increase in direct proportion to the weight classes. Thus, as the weight classes increase, the wrestlers are taller, broader, and have longer trunk and limb lengths.

Q. *Boxing*

The boxers are the second largest group after track (men). They have been divided into five weight categories for analysis: light fly and fly, bantam and feather, light and light–welter, welter and light–middle, and middle and heavyweight (see Tables 3.65, 3.75–3.81; Figure 3.12).

Age. There appears to be a slight increase in age from the two lightest to the heaviest weight categories. The oldest boxer is in the heavyweight category.

Somatotype. There is a trend toward increasing endomorphy and mesomorphy with increase in weight, and decreasing ectomorphy with increase in weight. In endomorphy the middle–heavyweight category is more endomorphic than the other groups. The welter–light–middle category is more endomorphic than the two lightest groups. In mesomorphy, the middle–heavyweight group is the most mesomorphic, and the two lightest weight categories are least mesomorphic, with the remaining two groups being mid-way between. The most ectomorphic groups are the two lightest weight groups, and the least ectomorphic is the middle–heavy group. The light–light–welter weight group is more ectomor-

Figure 3.12 Featherweight boxer, Mestizo, Mexico, age 21, somatotype 2–5–3½; mean for featherweight boxers, 1.7–4.9–3.

phic than the middle–heavyweight group, and less ectomorphic than the two lightest groups. Fifty-nine percent of the boxers are dominant mesomorphs, and 32% are in the ecto–mesomorphic category. The eight endo–mesomorphs all occur in the heavier weight categories. In the light–fly through featherweight categories, there is only one athlete who is on the endomorphic side of the mesomorphic axis of the somatochart, and his somatotype is a 2–6½–1. The highest endomorphic rating in the three lowest weight groups is 3 units, while the ectomorphic ratings range from 1 to 6, and mesomorphy from 3 to 7½. The five medalists in the light–fly and flyweight classes are very close to the center of their distribution. In the featherweight class, the gold medalist, Roldan, is slightly more ectomorphic than the majority of his group, but well within the distribution. The three medalists in the light and light–welter group are more mesomorphic than the medalists in the two lighter groups, but are still within their respective distributions. In the welterweight group, the silver medalist, Bessala, is the most mesomorphic in this group with a somatotype of 2–7½–1½. Of the two medalists from previous Olympics, one is in the center of the distribution, and the other is the most ectomorphic, with a somatotype of 1½–4½–4. In the middle and heavyweight class, two of the bronze medalists, Zaragoza, in the middleweight, and Rocha in the heavyweight, have somatotypes of 2–4½–3 and 2½–4½–2½, respectively. These somatotypes are somewhat less mesomorphic than the group, which has a mean of 6.0 and includes the most mesomorphic of the boxers; however, the bronze medalist in light–heavyweight, Dragan, has a 2½–6½–2 somatotype.

Body size. As with the wrestlers, no analysis of differences was made on the basis of weight. From Table 3.81 we see that height increases proportionally to the weight classes, with each weight class being significantly different from the others. The same pattern is seen for shoulder breadth. Although the ranking from the lightest to the heaviest weights is maintained for hip breadth, and the length dimensions, there are some between-group differences which are not significant. In hip breadth, the light and light–welterweight group, and the welter and light–middleweight groups are not significantly different. In arm and leg length, the two lightest groups do not differ, but the lightest group differs from the light and light–welterweight group. In trunk length, the two heaviest groups have a greater trunk length than the two lightest groups, and the heavyweight group differs from the light and light–welterweight group.

R. *Weight Lifting*

The weight lifters are divided into three combined weight groups—bantam and feather, light and middle, and heavyweight (including

Figure 3.13 Heavyweight weight lifter, Caucasoid, Belgium, somatotype $4\frac{1}{2}$–10–$\frac{1}{2}$; mean for light–heavy, middle–heavy, and heavyweight weight lifters, $3\frac{1}{2}$–7.7–0.7.

light–heavy, middle–heavy, and heavy) (see Tables 3.65, 3.82–3.86; Figure 3.13).

Age. All three groups have 17- or 18-year-old lifters, and all have lifters over 40. The oldest lifter is in the heavyweight category.

Somatotype. As might be expected, the weight lifters are the most mesomorphic athletes. The two lightest weight categories average almost 7 in mesomorphy, but they do not differ from each other, and the heavyweight lifters are one full unit of mesomorphy higher than those in the lighter weights. On endomorphy the two lightest groups do not differ from each other—both are low, and the skinfold values for some of the bantam and featherweight lifters are as low as those of athletes in some of the other sports. Some of the heavier lifters are relatively lean, but the majority have large skinfolds. All of the weight lifters have very low height–weight ratios and ectomorphic ratings. The groups range from a ½ unit to 2 units for the lowest two weight groups, and from ½ to 1½ rating units for the heavyweight group. The heavyweight group is significantly lower than the remaining two groups on ectomorphy. Sixty percent of the lifters are dominant mesomorphs, with 38% endo–mesomorphs. The one ecto–mesomorph is in the bantamweight class. There is a clear trend towards increasing endomorphy and mesomorphy as weight increases. Almost three quarters of those in the heavier weight categories are outside the border of the somatochart in the extreme endomorphic mesomorphy area. All seven medalists in this sports group are 7 or above in mesomorphy. In the two lightest weight categories, the three medalists average close to the 1–7–1 somatotype. In the heaviest weight category, the light–heavyweight silver and bronze medalists, Belaiev, has a somatotype of 2–8–½, and Ozimek has a somatotype of 1–7½–1½. Talts, the silver medalist in middle–heavyweight, has a 3–7½–½ somatotype. In the heavyweight, Reding the silver medalist is a 4½–10–½. This was the highest mesomorphic rating of any of the athletes. Another weight lifter has a rating of 7–9½–½.

Body size. Again, no weight analysis was carried out on the weight lifters. The strong relationship between the weight categories and the dimensions of height, hip breadth, and arm, leg, and trunk length are seen in Table 3.81. The three weight groups are significantly different from each other, and the values increase from the lightest to the heaviest weights. For shoulder breadth, the bantam–feather and light–middleweight groupings do not differ significantly, but the heavyweight group differs significantly from the other two.

S. *Reference Group (Men)*

This reference group of men taken from the Mexican population are to serve primarily as a point of comparison for certain measurements (see Table 3.87). They are not a true control group for the athletes from all over the world. A similar worldwide sample of non-Olympic competitors would be required. The results obtained probably reflect the racial basis of the particular reference group measured here.

Most subjects are relatively young and comparable to the Olympic males, as may be seen from their standard deviation, which is slightly under 5 years. With respect to height and weight the sample is quite variable, although not over the same absolute values as the Olympic sample. The mean somatotype for the reference group is $3\frac{1}{2}$–$4\frac{1}{2}$–3, a somatotype which is a central and slightly to the "northwest" of the center point of the somatochart in the endo–mesomorphy sector. The range of component values is quite wide on all three components. The skinfold values have a slightly positively skewed distribution.

T. *Reference Group (Women)*

The same caution should be used in interpreting the data on females as was indicated in the previous paragraph for the males (see Table 3.88). The age range is similar to that of the Olympic group with the exception of a few older women in the Olympic group. The heights and weights tend to be slightly lower than the sportswomen. The mean somatotype is in the meso–endomorphic sector of the somatochart, almost "due west" of the center. The high value of one subject on mesomorphy—that of 7.5—appears to be spurious, although there is no way of checking that at this time; therefore, this particular value should be viewed with caution.

U. *Summary*

In Section I we have described each of the sports and the subgroups within the sports, as well as making comparisons between the subgroups where there was more than one group within a sport.

Age

Although age is not a structural variable, it provides useful descriptive information on any population or sample. We have noted that the age range in many of the sports is quite considerable—in fact, almost four decades; but that within most of the sports the mean ages of the various event groups are fairly consistent.

Somatotype

Somatotype gives us a look at the form and composition regardless of absolute size. All of the male athlete groups show mesomorphy as the dominant component, but within and between the sports there are variations in the amount of ectomorphy and endomorphy, so that we find significant differences within sports in terms of somatotype. For the women, the somatotypes tend to be much more central than with the men, but most groups also show a slight dominance in mesomorphy, with ectomorphy or endomorphy also varying considerably within and between groups. The very obese and those very low in mesomorphy, are absent from both the male and female samples.

Body Size

The body size of wrestlers, boxers, and weight lifters is directly related to the weight category in which the participants compete. Other sports, except for rowing, cycling, and swimming for both men and women show considerable differences in body dimensions between groups. The rowers and cyclists are particularly homogeneous in both somatotype and body-size dimensions. The swimmers are also fairly uniform, but not quite as much as the rowers and cyclists.

We may summarize at this point by stating that there are some marked differences in somatotype and size between events in many of the sports we have examined. It is important to remember these differences when comparing different sports.

II. Analysis between Sports

The differences between sports are presented in this section. For the analysis of variance and multiple-range tests, the male athletes were arranged into 11 groups, and the female athletes into 6 groups. Comparisons were made between male sports groups containing no subgroups—gymnastics, modern pentathlon, basketball, water polo, and diving—and selected homogeneous sports groups. Because of the similarity of competitors in rowing, cycling, and swimming events, the athletes in these respective sports groups were combined. The kayak and Canadian canoeists were significantly different on 5 of the 10 variables. Therefore, for this analysis we used the kayak group as representing canoeists because it had the largest number of subjects ($N = 34$). Sprinters and a combined group of long-distance and marathon runners were chosen as the two groups to represent track and field. The sprinters differed from the distance runners on 8 of the 10 variables, while the two groups of distance runners did not differ significantly from each other on 9 of the 10 variables, but differed from most of the other track groups. None of the groups from the field events in Section I were used in this comparative analysis.

For the women, the gymnasts, divers, canoeists, swimmers, sprinters, and weight throwers made up the six groups. The first three groups contained no subgroups, and the within-sport analysis for the swimmers showed very few differences between the stokes. Sprinters and weight throwers were chosen to represent two distinct aspects of track and field.

The result of the analyses of variance and the multiple-range tests for the male and female athletes are presented in Tables 3.89 and 3.90. Before the analyses on all of the variables were conducted, subjects with missing data on any of the variables were removed.

In addition to the comparative analysis, Figure 3.14 shows the ranked absolute values by sex of the sports groups for age, height, weight, the three somatotype components, shoulder and hip breadth, and arm, leg, and trunk length. For all variables except age the values were obtained from the analysis of variance of the between-sport differences. Other values for groups in the figures and referred to in the text are from the descriptive data tables. The male and female reference groups are also shown in Figure 3.14 for comparison. In Figures 3.15 and 3.16 are presented distributions of the mean somatotypes for both male and female sports groups and reference groups.

A. *Age*

1. *Men*

As shown in Figure 3.14a, the mean ages for the male groups of athletes range from swimmers, who are the youngest, to long-distance runners, who are the oldest. However, weight throwers and weight lifters (see Tables 3.13 and 3.82) are almost 1 year older than the long-distance runner. Most of the groups average between 23 and 25 years. The reference group is younger than all other groups except the swimmers and divers. Of the four oldest groups, the weight lifters and weight throwers emphasize short-term speed and strength; the long-distance runners require great cardiovascular endurance; and wrestlers require strength and endurance (it is assumed that all sports require some degree of skill). It is commonly thought that the best endurance performances are achieved by older athletes, but in addition we see that competitions requiring a high degree of strength are also in the oldest age bracket.

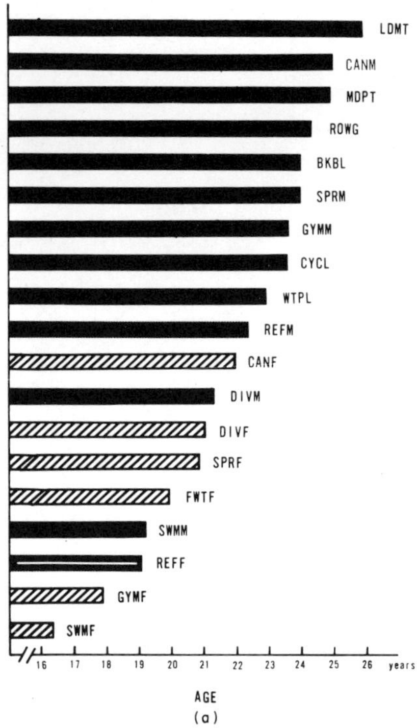

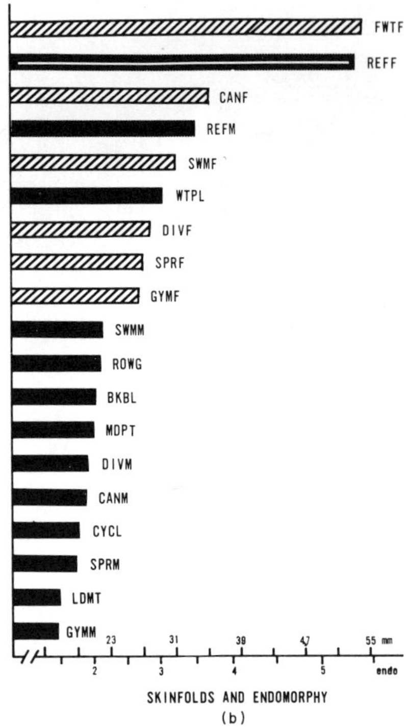

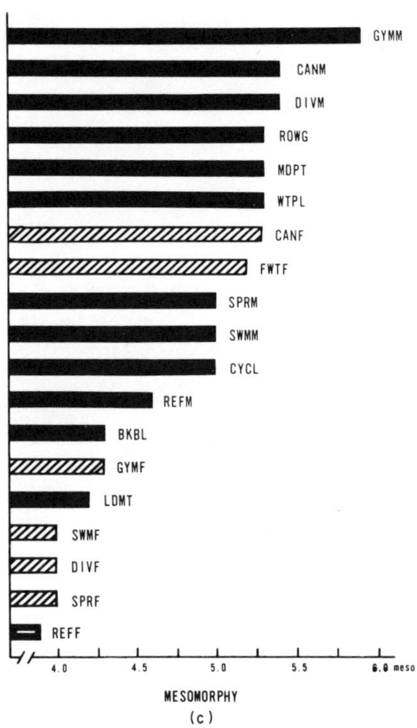

Figure 3.14 Means of variables indicated (Parts a–k) for males (black bars) and females (slanted hatching); reference groups (vertical hatching) are plotted for comparison. Abbreviations: LDMT, long-distance and marathon runners; CANM, male kayak canoeists; MDPT, modern pentathlonist; ROWG, rowers; BKBL, basketball players; SPRM, male sprinters; GYMM, male gymnasts; CYCL, cyclists; WTPL, water poloists; REFM, reference men; CANF, female canoeists; DIVM, male divers; DIVF, female divers; SPRF, female sprinters; FWTF, female weight throwers; SWMM, male swimmers; REFF, reference females; GYMF, female gymnasts; SWMF, female swimmers. (Continued on the following two pages.)

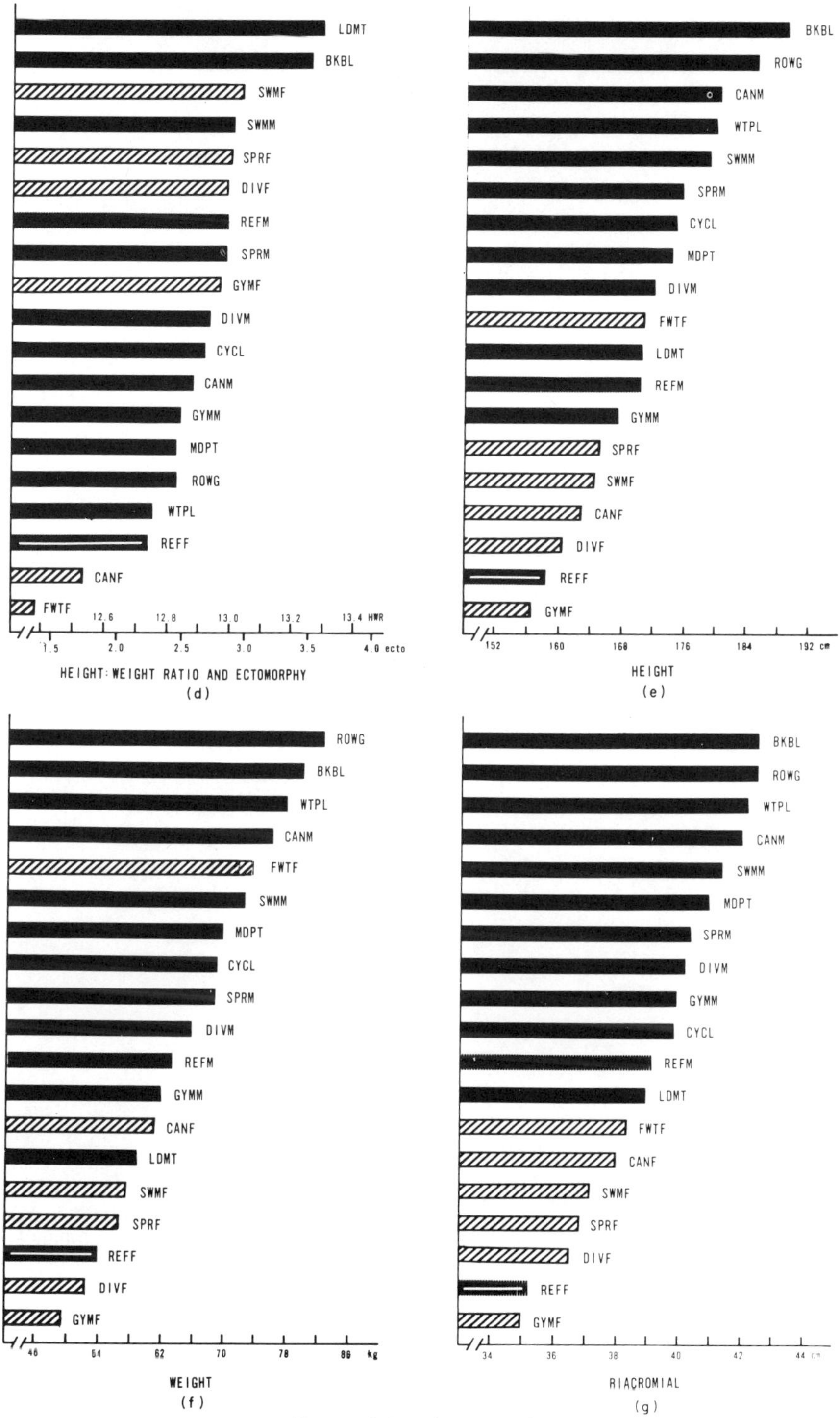

Figure 3.14 (continued).

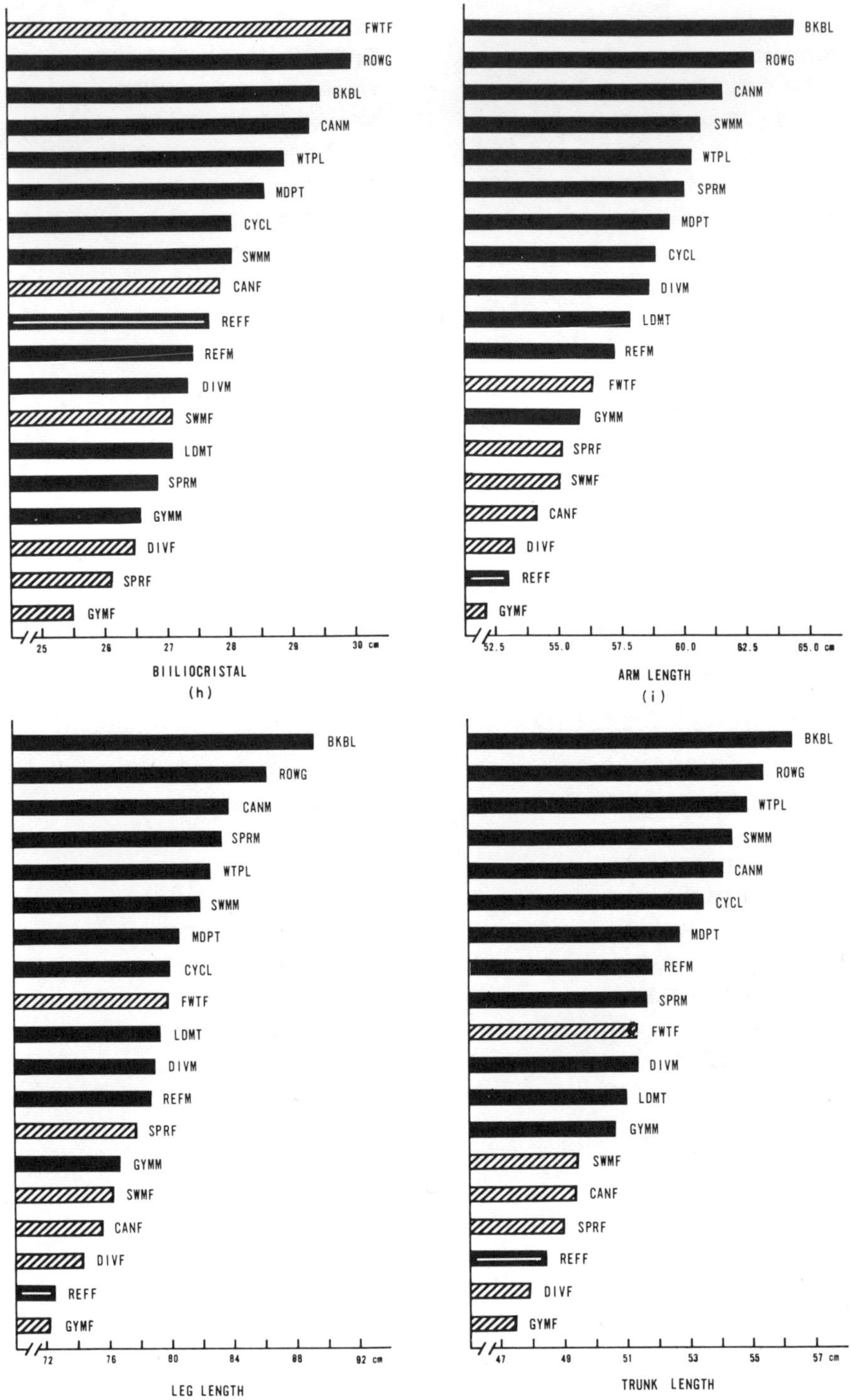

Figure 3.14 (continued).

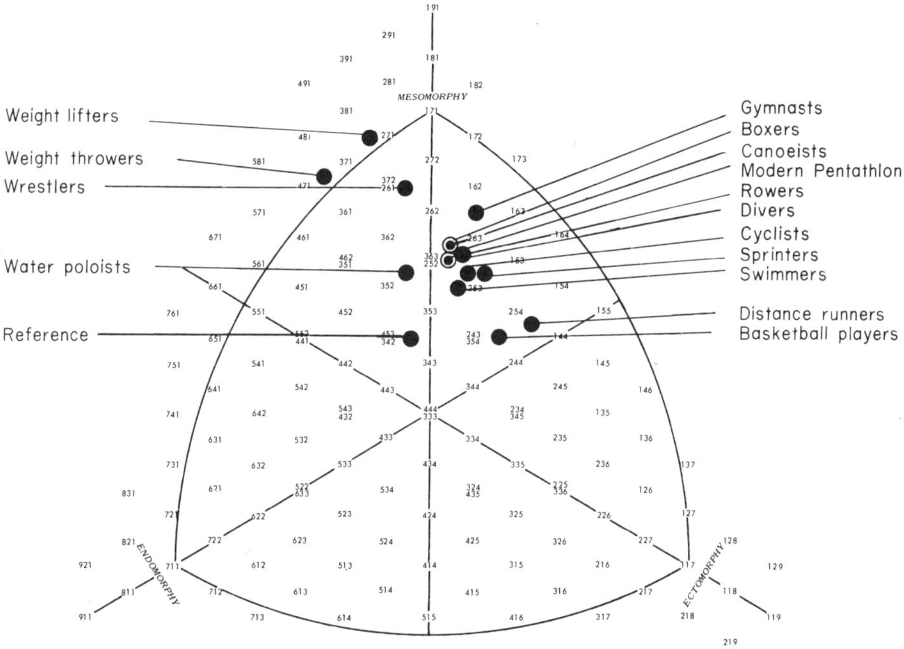

Figure 3.15 Somatotype distribution of mean somatotypes for various male sports groups and the reference group.

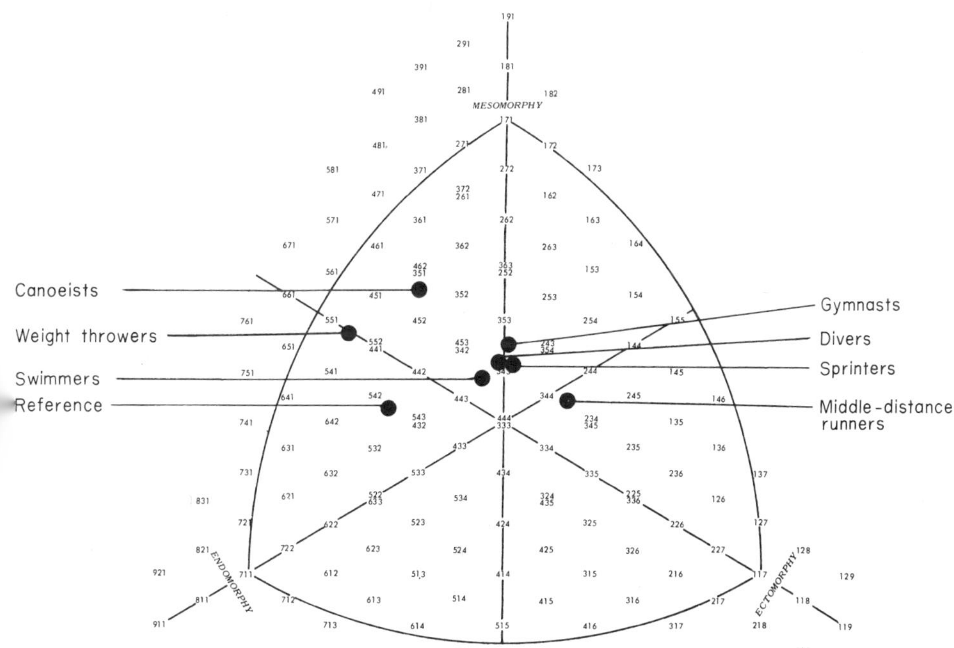

Figure 3.16 Somatotype distribution of mean somatotypes for various female sports groups and the reference group.

2. *Women*

All of the female groups are younger than the male groups except for the male swimmers. The female swimmers are the youngest, and the divers the oldest; however, divers, track and field, and canoeists are the oldest three groups at 20 to 21 years. The reference group is slightly younger than the majority of the women athletes. Only the gymnasts, who average close to 18 years, and the swimmers, who average a little over 16 years, are younger.

B. *Somatotype (See Tables 3.89 and 3.90; Figures 3.14b,c,d, 3.15, 3.16)*

1. *Endomorphy*

Men. The analysis of variance for the first component was run both with and without the water-polo players. The large variance of the water poloists indicated that they should not be included in the analysis; however, both *F*-ratios were significant, and almost identical conclusions were reached. In Table 3.89 both analyses are presented. The gymnasts and distance runners are significantly less endomorphic than the other groups, and sprinters and cyclists are less endomorphic than the basketball players, rowers, and swimmers. The water-polo players are significantly more endomorphic than the other groups. In addition, the reference males and the weight throwers are more endomorphic than the water-polo players. The majority of the male means range between $1\frac{1}{2}$ and $2\frac{1}{2}$ units of endomorphy (16 to 25 millimeters skinfolds).

Women. The analysis of variance for the women was run both with and without the weight throwers. The weight throwers had a very large variance and were withheld from the second analysis. Both analyses show that the weight throwers differ significantly from the remaining five groups (gymnasts, divers, sprinters, swimmers, and canoeists), who did not differ from each other. These groups average close to three in endomorphy. The weight throwers and reference women average almost $5\frac{1}{2}$ units in endomorphy and are clearly separate from the remainder of the groups.

2. *Mesomorphy*

Men. Gymnasts are the most mesomorphic of the male athletes. Seven of the male groups are concentrated in a narrow portion between the 5 and $5\frac{1}{2}$ units in mesomorphy, while the lowest groups are the reference males, basketball players, and long-distance runners. Table 3.89 shows that the gymnasts are significantly more mesomorphic than all of the groups except divers and canoeists; canoeists are significantly more mesomorphic than distance runners, basketball players, and cyclists; and the distance runners and the basketball players are less mesomorphic than the remaining groups.

Women. The women are grouped closely around the 4 unit in mesomorphy, with the sprinters, divers, swimmers, and gymnasts not differing significantly from each other. The weight throwers and canoeists are the most mesomorphic of the women's groups and differ significantly from the other four groups. The reference group of women is close to the values for the sprinters, swimmers, and divers.

3. *Ectomorphy*

Men. On the third component, ectomorphy, the long-distance runners and basketball players are highest. Most of the remaining male groups are between 2½ and 3 in ectomorphy, including the reference group. Table 3.89 shows that the distance runners and basketball players are significantly more linear than the remaining nine groups, and the swimmers and sprinters are more linear than the rowers and water-polo players.

Women. The female gymnasts, divers, sprinters, and swimmers average close to 3 units in ectomorphy and are significantly more ectomorphic than the canoeists or weight throwers. The reference women are in between the weight throwers and canoeists and the other four groups.

4. *Somatochart Distribution*

Men. Figure 3.15 shows that the means of several groups of athletes (boxers, canoeists, modern pentathlon, rowers, field events, divers, cyclists, sprinters, and swimmers) are contained in a small area bounded by the 2–5–3, 3–6–3, 2–6–3 somatotypes. The average somatotype describing all of these groups is a 2–5½–3. The remaining male groups differ with respect to the amount of endomorphy and ectomorphy, and some differ with respect to mesomorphy. The most endo–mesomorphic athletes are the weight throwers and the weight lifters. Both groups have a mean close to 7 in mesomorphy and 2½ or 3½ in endomorphy. They are the least linear of the athletes. Compared to the weight throwers and the weight lifters, the wrestlers are slightly less endomorphic and mesomorphic, and are slightly more ectomorphic. The water poloists are endo–mesomorphic also, but are not quite as mesomorphic as the weight throwers and lifters and wrestlers. The gymnasts are ecto–mesomorphs, with an average close to 6 in mesomorphy, but they have very lean physiques with low skinfolds. The most ectomorphic are the distance runners and basketball players, whose mesomorphy drops with the rising ectomorphy. The reference male somatotype is near the center of the somatochart and is more endomorphic and less mesomorphic than the other male groups.

Women. The somatotypes of the female athletes from gymnastics, sprinting, diving, and swimming are centered very close to the 3–4–3 somatotype. The canoeists and weight throwers are endo–mesomorphic, with mesomorphy dominant in the canoeists, and endomorphy and mesomorphy being at the same level for the weight throwers. The reference women are meso–endomorphic, with endomorphy dominating over mesomorphy.

C. *Body Size* (*See Tables 3.89 and 3.90*)

1. *Height*

Men. From Figure 3.14e we see that there is a range of group means of almost 22 centimeters between the shortest male athletes, the gymnasts, to the tallest, the basketball players. The rowers are 4 centimeters shorter than the basketball players. From Table 3.89 we see that the gymnasts are significantly shorter than all other groups, and the rowers and basketball players are significantly taller, as well as being different from each other. At the intermediate heights, canoeists, water-polo players, and swimmers are taller than sprinters, cyclists, modern pentathlon, divers, and distance runners. Distance runners and divers do not differ from each other, but the distance runners are shorter than the pentathlon, cyclists, and sprinters. The reference men are shorter than all groups except the gymnasts.

Women. The weight throwers are significantly taller than all other groups and the gymnasts are significantly shorter than the swimmers or sprinters. The reference women are approximately 1 centimeter taller than the gymnasts and, therefore, shorter than the remaining women's groups.

2. *Weight* (*See Tables 3.89 and 3.90; Figure 3.14f*)

Men. The distance runners and gymnasts do not differ from each other and are the lightest athletes, while the rowers are significantly heavier than all other groups. The basketball players are significantly heavier than all other groups except the rowers and water-polo players. The basketball players, water-polo players, and canoeists, are heavier than the swimmers, pentathlon, cyclists, sprinters, or divers; and in addition, the swimmers are heavier than the pentathlon, cyclists, sprinters, and divers. The reference group is lighter than all groups except the gymnasts and distance runners.

Women. Because of the large variance in the weight-throwing group, the analysis of variance was run both with and without this group. Both analyses produced the same result, indicating that the weight throwers were significantly heavier than the remainder of the groups, and the gymnasts were significantly lighter than the sprinters, swimmers, and canoeists. The female weight throwers with a weight of more than 73 kilograms are heavier than most of the male groups and are approximately 23 kilograms heavier than the lightest female group, the gymnasts. The reference population is lighter than all groups except the divers and gymnasts.

3. *Biacromial Breadth* (*See Tables 3.89 and 3.90; Figure 3.14g*)

Men. The basketball players, rowers, and water-polo players have significantly broader shoulders than the distance runners and the cyclists. The groups between the gymnasts and canoeists seem to overlap considerably and no clear pattern of difference is apparent. A large number of groups have means between 40 and 41 centimeters, but the reference group has the narrowest shoulders of all the groups except the distance runners.

Women. The gymnasts have significantly narrower shoulders than all other groups except the divers. The reference group has the same shoulder dimension as the gymnasts.

4. *Biiliocristal Breadth* (*See Tables 3.89 and 3.90; Figure 3.14h*)

Men. The rowers and basketball players have the widest hips except for the canoeists, while the narrowist hips are found in the gymnasts, sprinters, and distance runners. The reference group is about the center of the groups.

Women. As women have a relatively broad pelvis, we would expect to find a greater tendency to overlap with the male distribution in this dimension. The weight throwers rank with the male rowers in terms of absolute hip breadth. The female canoeists, reference women, and swimmers, in addition to the weight throwers, exceed the hip breadth of the male gymnasts. Among the women, the weight throwers have significantly larger hip breadth than all other groups. The gymnasts are significantly smaller in hip breadth than the swimmers and canoeists.

5. *Arm Length* (*See Tables 3.89 and 3.90; Figure 3.14i*)

Men. The basketball players have the longest arms, closely followed by the rowers; and gymnasts have the shortest arms. The basketball players' arms are significantly longer than the rowers, and both of these are significantly longer than all of the groups except the canoeists. The gymnasts' arms are significantly shorter than all groups, and the distance-runners' arms are shorter than all except the divers' arms. In addition, swimmers and water-polo players have longer arms than cyclists and divers.

Women. The weight throwers have significantly longer arms than gymnasts and divers, and in addition, the sprinters and swimmers have longer arms than the gymnasts. The range of values for the women is fairly small. The reference women have shorter arms than all groups except the gymnasts.

6. *Leg Length* (*See Tables 3.89 and 3.90; Figure 3.14j*)

Men. The longest legs belong to the basketball players, with rowers' about 3 centimeters shorter. The shortest legs are those of the gymnasts. The reference men have shorter legs than all except the gymnasts. The difference in leg length between the rowers and basketball players is significant, and these two groups differ significantly from the remaining groups. Gymnasts have shorter legs than all groups except the divers, while canoeists, sprinters, water-polo players, and swimmers have longer legs than distance runners and cyclists.

Women. The same pattern of differences is seen for leg length as for arm length. The weight throwers have longer legs than the gymnasts or divers. Sprinters and swimmers have longer legs than gymnasts. The leg length of the reference women is almost the same as that of the gymnasts, and shorter than that of all of the other groups.

7. *Trunk Length* (*See Tables 3.89 and 3.90; Figure 3.14k*)

Men. The basketball players and rowers have the longest trunks, while the gymnasts have the shortest trunks. The reference group averages close to 52 centimeters, similar to that of the sprinters and divers. The gymnasts and distance runners have significantly shorter trunks than canoeists, cyclists, swimmers, water-polo players, rowers, or basketball players. Divers, sprinters, and pentathlon competitors have significantly shorter trunks than water-polo players, rowers, or basketball players. The latter three have significantly longer trunks than the other groups, with basketball players having longer trunks than water-polo players.

Women. Weight throwers have significantly longer trunks than gymnasts, divers, or sprinters; and swimmers have significantly longer trunks than gymnasts. The reference group has a longer trunk than either the divers, or the gymnasts, but shorter than the other group.

D. *Summary*

The comparison between sports shows that there are marked differences in body size and somatotype. The greatest differences are likely to be due to the requirements of the different events. On the other hand there are few differences between some of the sports indicating that the same physiques can excel at different events and that such excellence probably depends on sociocultural and personality factors rather than on physique. These findings are similar for both male and female athletes.

III. Analysis by Race

Sections I and II have shown that there are similarities and differences within and among some sports. A question which readily comes to mind is whether or not these similarities and differences are affected by race. In view of the fact that people from different parts of the world and from different racial stock appear to differ on selected anthropometric characteristics (Martin and Saller 1959), one would expect the magnitude of these differences to be reflected in the Olympic athletes. On the other hand, it may well be that the requirements of the event at the Olympic level of competition select out those from the various countries and races who have the attributes which are suitable for successful participation in certain events. We shall attempt to answer these questions by examining the differences between races within events.

Table 1.2 shows the number of Olympic athletes by sex and race. Tables 3.91 and 3.92 show the racial distribution of male and female athletes by event and sport as well as showing the frequencies for the reference groups. For the racial analysis, the subjects who were not classified according to race (unknown) were removed and those with Negroid characteristics were grouped into one category. The three remaining categories were Caucasoid, Mongoloid, and Mestizo. The frequencies in these two tables were used as the bases for deciding which groups to compare within events and sports. In some sports, one group predominated (usually Caucasoid), with the numbers so small in the other categories as to make any comparisons meaningless. In other events or sports, comparisons were made between two, three, or four racial groups, depending on the frequencies available for the particular event or sport. Further information on the bases for selection will be given in considering each of the athletic groups. For purposes of comparison, descriptive analyses for male and female athletes and reference groups by race are given in Tables 3.93–3.103. These descriptive statistics can be used to compare the values for the event and sport groups by race, but the reader is reminded that the subjects themselves have contributed to the means they are being compared with. The values for the reference groups are useful as they give us some idea of the differences which might be expected between Caucasoid and Mestizo men and women, who are not selected for athletic ability. Before the analyses of variance were run on the various racial groups, the subjects for the groups with very small numbers and the unknown racial categories as well as those who have missing data on any of the 10 variables chosen for analysis were removed. The means, standard deviations, F-ratios, and multiple-range test results are presented in Tables 3.104–3.125. The results will be presented in the same order as for the analysis within sports.

A. *Track and Field*

1. *Men*

Sprinters. The Mestizo sprinters are significantly shorter in stature, arm length, and leg length than the Negroid athletes (see Table 3.104). The Mongoloid and Negroid sprinters have narrower hips than the Caucasoid sprinters, and the Mongoloids have shorter arms than the Negroids. None of the groups differ significantly in weight, shoulder breadth, trunk length, or the three somatotype components.

Four hundred meters. Three groups of the 400-meter runners were compared as there were no Mongoloid athletes in this sample (see Table 3.105). The Mestizo athletes are shorter than the Caucasoid and have shorter arms than the Negroids, and shorter legs than either Caucasoids or Negroids. The Negroids have narrower hips than the Mestizos or Caucasoids. The groups do not differ in body weight, shoulder breadth, trunk length, or in the three somatotype components. The pattern of differences is very similar among the sprinters and the 400-meter runners.

Middle distance. The four Mestizos were dropped from this group and the comparison made between the Caucasoid and Negroid athletes (see Table 3.106). In contrast to the two previous running events, the Caucasoids predominate numerically over the Negroids. Although the two groups do not differ significantly in height, they differ significantly in hip width, in which the Negroids are smaller than the Caucasoids, and in arm and leg length, in which the Negroids are longer than the Caucasoids. These findings would indicate both absolute and relative differences between the two groups. There are no significant differences on the other seven variables.

Long distance and marathon. The long-distance and marathon runners were combined to form a single group (as for the between-sport comparisons) so that all four racial categories could be compared (see Table 3.107). There are a great variety of differences between the racial groups. They differ on 8 of the 10 variables and are the only event group with any major differences on somatotype components. The two variables on which there are no significant differences are body weight and shoulder breadth. Although not all patterns are consistent, two groups emerge as being similar to each other and, to varying degrees, different from the other two groups. The Mestizos and Mongoloids are the smallest of the four groups, while the Negroids and Caucasoids are the largest. The exception to this is on hip width, where the Caucasoids have larger hips than the Mongoloids or Negroids and do not differ from the Mestizos. On somatotype, the main difference appears to be caused by the Mestizos, who are more endomorphic than the other three groups, more mesomorphic than the Caucasoids or Negroids, and less ectomorphic than the Caucasoids and Negroids. In addition, the Mongoloids are more mesomorphic than the Negroids and less ectomorphic than the Caucasoids and Negroids.

Walkers. The Caucasoids and Mestizos are compared in this group. The Caucasoids are taller and heavier and have longer legs than the Mestizos, but otherwise there are no significant differences between the two groups (see Table 3.108).

Jumpers. The single Mestizo jumper was dropped and the remaining three racial groups were analyzed (see Table 3.109). Again, the number of Negroids in the sample predominate over the Caucasoid and the Mongoloids. There is a definite pattern of body size exhibited in the analysis. The Mongoloids are the smallest and the Caucasoids are the largest. All three groups differ significantly from each other in height, and the Mongoloids are significantly lighter, have narrower shoulders, shorter arms and legs, than the Negroid and Caucasoid athletes. On the other hand, the Caucasoids are more ectomorphic than the Negroids, have wider hips and longer trunks than either the Mongoloids or Negroids. Although they differ in body size, the three groups show no difference in the somatotype components, indicating that their body form and composition is somewhat similar.

2. *Women*

Sprinters. The Mestizo, Negroid, and Caucasoid athletes differ significantly from each other in height, and the Caucasoids are heavier than the Mestizos in weight (see Table 3.110). Although there are no differences between the groups in shoulder breadth, the Negroids have significantly narrower hips than the other two groups and have longer arms and legs than the Mestizos. The Caucasoids have significantly longer trunks than either Negroids or Mestizos. Although the F-ratio is not significant, the multiple-range test indicates a possible difference between the Negroids (low) and Mestizos (high) on skinfolds; otherwise, there are no differences between the groups on somatotype components.

Middle distance. The comparison for the middle-distance runners was made on the Caucasoid and Negroid groups (see Table 3.111). The Caucasoids are significantly heavier, have wider shoulders, wider hips, and longer trunk lengths than the Negroids. There are no significant differences in height, arm or leg length, nor in somatotype.

Jumpers. With the relatively small samples of both groups, it was not surprising to find no significant differences between the Caucasoid and Negroid jumpers on 9 out of the 10 variables (see Table 3.112). The Negroid jumpers have significantly narrower hips than the Caucasoid jumpers.

B. *Swimming*

1. *Men*

Because of the small numbers and similar proportions of Caucasoids and Mestizos through the five swimming events, and because very few differences had been shown between events, the male swimmers were considered as a total group with two racial categories, Caucasoid and Mestizo (see Table 3.113). The Caucasoids and Mestizos do not differ from each other in gross body size, lengths, breadths, or somatotype components. The single exception is the shoulder breadth, in which the Caucasoids have wider shoulders than the Mestizos.

2. *Women*

As with the male swimmers, the females are divided into two racial groups, Caucasoid and Mestizo (see Table 3.114). There are no differences between the two groups on body size, lengths, breadths, or somatotype components. For the analysis of the skinfolds, the subject referred to in Sections I and II who had a high skinfold total was removed to eliminate the skewness of the distribution.

C. *Basketball*

The two predominant groups in basketball were Caucasoid and Negroid. The analyses of variance show that the Caucasoid basketball players are significantly heavier, have wider hips and longer trunk lengths than the Negroids (see Table 3.115). There are no differences in height, shoulder width, arm and leg length, and somatotype components. The absolute values of the differences between the means are quite noteworthy for this group. There are differences of approximately 8 kilograms in weight and slightly more than 3 centimeters in hip breadth and trunk length.

D. *Rowing*

The predominant group in rowing was Caucasoid ($N = 71$), but there were sufficient Mestizos ($N = 13$) to warrant a comparison. The analysis shows that the Mestizos are significantly shorter, lighter, have narrower shoulders and shorter arms and legs than the Caucasoids (see Table 3.116). They do not differ on somatotype, hip width, or trunk length. As was shown in Table 3.52, there were no differences between the various styles of rowing within the sport, so this racial difference takes on added significance when considering the performance of the crews.

E. *Cycling*

Four racial groups were analyzed in cycling, although the number of Negroid performers was low ($N = 7$) (Table 3.117). The analyses show that the significant differences are in terms of gross body size, breadths and lengths, whereas there are no differences in the somatotype components. In general, the Mongoloids and Mestizos are the smallest cyclists and the Negroids and Caucasoids are the largest. The two exceptions to this are the Negroids, who had the narrowest hips and the shortest trunk length and differed significantly from Mestizos and Caucasoids in the former, and the Caucasoids in the latter. The Caucasoids are taller and heavier than the Mongoloids and Mestizos, have broader shoulders, and longer arms and legs than the Mongoloids. They have significantly longer trunk lengths than the Mestizos.

As with the rowers, the cyclists did not show significant differences between the events, but among the cyclists there are significant differences in terms of race.

F. *Gymnastics*

1. *Men*

The analysis was carried out on Caucasoid, Mestizo, and Mongoloid gymnasts, although the last group had only three subjects (see Table 3.118). Except for hip width, in which the Mongoloid gymnasts are narrower than the other two groups, there are no significant differences between the groups on any of the variables. In terms of ranking only, there is a suggestion that the Mongoloids are the smallest, the Mestizos next largest, and the Caucasoids the largest of the three groups.

2. *Women*

The Caucasoïd athletes differ from the Mestizos in that they are taller, heavier, have wider shoulders and longer trunks than the Mestizo athletes (see Table 3.119). They do not differ on hip width, arm or leg length, or on somatotype components.

G. *Freestyle Wrestling*

As can be seen from Table 3.91, the overwhelming majority of Greco–Roman wrestlers are Caucasoids, but in freestyle wrestling there is a distribution through the three other race categories, although Caucasoid still predominates. An examination of the distribution of races by weights shows that there are too few in the heavyweight category (six Caucasoid, one Mongoloid) to be useful in the analysis; consequently, only freestyle wrestlers and flyweight through middleweight were used for the analysis. Theoretically, if there is a proportional distribution of races through the weight categories, then there should be no significant differences on weight. Table 3.120 shows that except for two variables there are no significant differences between any of the four racial categories. Regarding hip width, the Negroids are narrower than the Mestizos or Caucasoids, and on trunk length the Negroids are shorter than the Mongoloids. On the other dimensions and somatotype there are no significant differences between the groups.

H. *Boxing*

The boxers were divided into two weight categories for the racial analysis (see Tables 3.121 and 3.122). In Group (a) were placed the light–fly through lightweight boxers who had a reasonably similar distribution of the four races through the weight categories. A second group (b) was formed using three race categories for the welter and light–middleweights, with the elimination of Mongoloid category as there were no athletes in this division. Furthermore, the heavyweight boxers were eliminated from analysis partly because of wide divergence in sizes, which would unduly influence the remainder of the groups.

1. *Light–Fly through Lightweight Boxers*

The four racial groups do not differ significantly with respect to weight, nor did they differ in height, shoulder breadth, or arm length, nor the somatotype components. The Negroids are significantly narrower in hip breadth, and longer in leg length, than the other three groups, while in trunk length they were significantly shorter than the Caucasoids or Mongoloids. The difference between the Mestizos and the Negroids in height should be regarded with suspicion because of the insignificant F-ratio.

2. *Welter and Light–Middleweight Boxers*

The Caucasoid, Mestizo, and Negroid boxers at these weights do not differ significantly in terms of body size, length and breadth dimensions, or somatotype, except for the Negroid boxers who have a significantly narrower hip breadth than either Mestizo or Caucasoid boxers.

I. *Weight Lifting*

As with wrestlers and boxers, the heavy weight lifters were removed from analysis—the majority of these were Caucasian (see Table 3.123). At the two lighter weight groups used for within-sport analysis, bantam through middleweight, there were approximately equal proportions of Caucasoids and Mestizos, but there were slightly more Mongoloids at the lighter weights. However, the three groups were used for comparison, and the results do not seem to be unduly effected by the lighter Mongoloids. The analysis showed that there is no significant difference between the groups on any of the characteristics except hip width, in which the Mongoloids are significantly smaller than the Caucasoids. Overall, the Mongoloids tend to be the smaller group, while the Mestizos and Caucasoids are very similar to each other in most dimensions and in somatotype.

J. *Reference Groups*

1. *Men*

A comparison of Caucasoid and Mestizo men in the reference group shows that the Caucasoids are significantly taller, heavier, less mesomorphic, more ectomorphic, have longer arms, legs, and trunks than the Mestizos (see Table 3.124). They did not differ significantly in skinfolds, or hip or shoulder width.

2. *Women*

The Caucasoid women are taller, lower in endomorphy and mesomorphy, higher on ectomorphy, and have greater arm, leg, and trunk lengths than do the Mestizos (see Table 3.125). There are no significant differences in weight or shoulder and hip widths. In these two groups it is interesting to note that the difference in height for the same weight can be accounted for in the increased endomorphy and mesomorphy of the Mestizos; and the increased mesomorphy is perhaps also represented in the similarity of shoulder and hip width, in spite of the difference of 4 centimeters in height.

K. *Summary*

In summary, we see that the racial analysis shows that performers in the same event have similar somatotypes. Second, performers in the same event may differ in terms of gross body size, or in breadth and lengths of limbs or trunk. However, the most persistent of these differences is the narrow (both absolute and relative) hip breadth of the Negroid athlete. In general the athletes decrease in size from the Caucasoids, Negroids, Mestizos to Mongoloids. These patterns are true for the total samples of athletes, both male and female, as well as for the differences within sports or events.

IV. Discussion of Results

When considering the absolute values of the dimensions of the body, one would expect a fairly consistent relationship among them. This is shown in the between-sport analyses by the consistently high-ranking and absolute values for male basketball players and rowers, who have the greatest absolute size, and the low-ranking and absolute values for the gymnasts and distance runners, who have the least absolute size. Although the male weight throwers (shot, discus, and hammer) were not included in the between-sport analyses, they have higher absolute values for body weight, shoulder and hip breadth, and trunk length than the other 11 sport groups (see Table 3.13). They are second only to basketball players in height and arm and leg length; and they have the lowest HWR's of all sport groups.

For women, the weight throwers and canoeists appear at the high end of the distribution, and the gymnasts and divers at the low end. In general, both the male and female reference groups appear to be smaller than the athletes.

Although Tables 3.1, 3.10, 3.17, 3.26, 3.34, 3.53, 3.69, 3.75, and 3.82 show the absolute mean values of the variables for track and field, swimming, canoeing, wrestling, boxing, and weight lifting, the reader is reminded that these groups are a composite of significantly different events in terms of both somatotype and size. These mean values represent the mid-point in a very wide distribution on most variables. For example, with respect to weight, the mean of the wrestlers is 70.5 kilograms but some of the heavyweight wrestlers weigh over 100 kilograms. On the other hand, some of the wrestlers, weight lifters, and boxers are smaller than some of the gymnasts. Because somatotype is a rating of the relative body form and composition, it is not influenced by size. Again, the reader is cautioned, however, and should note that although some groups of athletes have the same somatotype, the groups differ considerably in body size. For instance, rowers, modern pentathlon competitors, and divers have similar somatotypes, but differ vastly in size.

For comparison with other studies it should be noted that the height used in this analysis is the maximum stretch height (stature), which tends to eliminate the diurnal variation in height. The values for maximum stretch height average 1–2 centimeters more than the values for freestanding height.

Some pairs of comparisons between selected sports may prove of value. Rowers and basketball players appear to be somewhat similar in characteristics. However, the basketball players are significantly taller and lighter than rowers. Along with this, they are significantly less mesomorphic and more ectomorphic, indicating that the difference in weight is primarily one of muscle and bone tissue. The two groups do not differ on shoulder or hip breadth, or on trunk length, but the basketball players have longer arms and legs.

Swimmers and water-polo players compete in the same medium and would appear to start from similar basic proficiency in the water. We find that although they do not differ in height, the water-polo players are significantly heavier and have higher skinfolds than the swimmers. Although they are equally mesomorphic, the greater skinfolds (endomorphy) appear to cause the significant difference in ectomorphy, in which the water-polo players are lower than the swimmers. Water-polo players have broader shoulders and hips and longer trunks than the swimmers, but they do not differ in arm length or leg length.

Some similarities are seen between cyclists and track sprinters. They are similar in height, weight, endomorphy, mesomorphy, ectomorphy, biacromial breadth, and arm length, but the sprinters have narrower hips and longer legs than the cyclists.

Divers are often considered to be a form of gymnast. Our analysis shows that this claim is reasonably true. Divers are taller than gymnasts and have higher skinfolds. On the other hand, the two groups do not differ in weight, mesomorphy, or ectomorphy, or in shoulder or hip width, or trunk and leg length. The only other significant difference is that the gymnasts have shorter arms. For the women, the similarity between the gymnasts and divers is also very great. Although the gymnasts are lower on all values except mesomorphy, the divers and gymnasts are ranked next to each other on 8 of the 10 variables. They do not differ in height, weight, endomorphy, mesomorphy, ectomorphy, shoulder or hip breadth, arm length, trunk length, or leg length.

We have noted the somatotype positions of some outstanding athletes in relation to their fellow competitors. Some are within or at the centers, and others are at the extremes of their distributions. There is no consistent pattern one way or the other. When physique is equated within an event, then other factors such as psychological condition, training technique, and tactics probably determine the differences in performance. When physiques are different and other factors are equated, then the physique factor alone may determine the performance differences. It must be remembered that, even at moderate levels of performance, structure alone does not determine performance. Although initial recognition must be given to physical structure, such recognition (as height for a potential basketball player) is soon superseded by certain performance criteria.

The variability of the champions and the other competitors in an event must not be overlooked, although this variability is not close to that found in a general population. Statistically speaking, the best estimate of the physical characteristic of a sample is the mean, but in practice the value of the characteristic for the individual is more important. From our analysis we can conclude that moderate levels of variation can occur among individuals in an event in which all have reached Olympic levels of performance. If the individual aspiring to this level of performance has, or can acquire by training, the physical characteristics within the variation shown in the event, then this part of his requirement for success is satisfied. In general we would expect better prediction for success the closer the individual is to the criterion. Given the prerequisite structure, efforts can then be concentrated on aspects of conditioning and the acquisition of skill. With regard to the variation in physical characteristics, Carter (1970) has shown that the pattern of successful physiques becomes narrower as the level of competition increases.

Where we observe that athletes with the same characteristics succeed in different events or sports, we may suppose that the acquisition of the different skills must be a question of such things as training, cultural–ethnic background, or socioeconomic opportunity. These latter factors may also explain the prominence of certain countries in some events or sports. For a further discussion see Jokl (1964) and Hirata (1966).

Although a detailed comparison of our results with those of previous investigators is not warranted at the moment, we can make some general observations. The trends in track and field (men) noted in earlier studies have been generally confirmed and in addition they have been made more precise by our larger samples and our comparative analysis. Completely new data have been presented for some events, sports, and for women athletes. A statistical comparison between different sports has demonstrated both similarities and differences in terms of physical structure.

As always, the findings must be interpreted with consideration for the limitations of the data. Although our samples in general are large and diverse enough, some subtle biases are probably present. In addition, some of our samples are very small. On the other hand, our statistical analyses have allowed for these small samples. Many readers will want to know why we did not analyze certain measurements or comparisons. For this reason, further analysis of selected combined measurements (ratios) is planned, but for the moment we are concerned with establishing the existence of certain general patterns of structure as related to function.

To summarize, we may state that in Sections I and II we have examined the body form and size of the athletes. We have described the prototypes and their variations for the different athletic functions which they perform. We have seen that the same prototypes excel at the same function, but in addition, the same prototypes can also excel at different functions. Finally, some functions may demand vastly different prototypes.

It is necessary to make some further cautionary remarks regarding the descriptive data for the male and female athletes by race. To begin with, the number of athletes by race may not reflect a true stratified sample by events. Second, there are many Caucasoids in the events requiring large athletes such as weight throwers, heavyweight wrestlers, boxers, weight lifters, rowers, and water-polo players. Third, the apparent differences between the races in these tables do not necessarily reflect true population differences. However, we have noted that in comparing the Caucasoids and Mestizos from a resident population in Mexico that there are significant differences between the two groups. By definition the Mestizos are mixtures and so are the Negroids in our sample—especially those from the United States of America. No accounting for the precise amount and origins of the mixtures is attempted in this analysis.

In general, the tables show that the athletes decrease in size from the Caucasoids, Negroids, Mestizos, to Mongoloids. This pattern is true for the total sample of athletes, both male and female, as well as for the differences within sports or events, except in the latter case where weight is controlled, then there are fewer differences in size. With reference to the two breadths and the three lengths, the most striking aspect of the analysis is that in 13 of the comparisons between the Negroids and Caucasoids, the Negroids are significantly narrower on hip breadth than are the Caucasoids—they differ less from the Mestizos and the Mongoloids. In contrast, Caucasoids are broader in hip and shoulder breadth than the Negroids and they differ less from the Mestizos and Mongoloids in these variables. Furthermore, the Negroids have longer arms, legs, and shorter trunks than some of the other three groups. These remarks refer to the absolute values. When relative values are studied (by dividing the mean dimension by mean stature) the Negroids are invariably narrower on hip breadth, longer on arm length, leg length, and shorter on trunk length when compared to the other three racial groups. These findings are in agreement with general anthropological statements on structural differences between races.

The F-ratios for the somatotype ratings in the various components are not significant in 19 of the 20 events or sports. The event in which the components are significantly different is long-distance and marathon running, but we should remember that these differences appear to be caused primarily by the Mestizo group being more endomorphic and mesomorphic, and less ectomorphic than the others. The difference on the first component, endomorphy, may be purely a reflection on the amount of training of these particular athletes, as the skinfold values do not appear to be different in the other events. With this single exception, it would appear that the same somatotypes excel at the specified events regardless of race, and that the functional requirements of the events demand similar somatotypes.

One difference in terms of somatotype among the racial groups for all the male athletes regardless of event is that the Negroid athletes differ from the Caucasoid, Mongoloid, and Mestizo in the mesomorphic and ectomorphic components. In mesomorphy, they are approximately half a unit lower (at 4.8), and half a unit higher (3.0) in the third component. Among the female athletes, the Mestizos appear to be higher on endomorphy and mesomorphy, and lower in ectomorphy, than either Caucasoids or Negroids, while the Negroids appear to be lower in endomorphy and mesomorphy than either Caucasoid or Mestizo. These somatotypic observations on Caucasoids and Mestizos are also seen in the female reference group. Again, because of the limitations of our stratified sampling, these results should be viewed as indications only and not viewed as being statistically significant.

In summary, we can see that the racial analysis shows that performers in the same event have similar somatotypes. Second, performers in the same events may differ in terms of gross body size, or in breadths and lengths of limb or trunk. However, the most persistent of these differences is the narrow (both relative and absolute) biiliac breadth of the Negroid athlete as well as longer arms and legs. These would appear to be true racial characteristics as they appear in virtually all the events.

There is one very important question which is not answered by these data. Do athletes from different countries (and sometimes, races) excel at certain events because there is a large proportion of the appropriate somatotypes and sizes available from which to select the athletes, or to put it differently, have some countries and races failed to excel at certain Olympic events because there are insufficient numbers of the appropriate size and somatotype from which to select or to train for the Olympics? The answers to these questions await further study.

V. Conclusions

In conclusion we can assert with some confidence that with respect to Olympic athletes:

1. There are significant differences in the size and somatotype of participants among events in some sports, but there are few or no differences among events in other sports.
2. There are significant differences in the body size and somatotype of participants in different sports, but some sports participants are similar in size and/or somatotype.
3. Performers with similar somatotypes excel at specific events regardless of race.
4. Performers in the same events from different races may differ in terms of body size.

Having pointed out these structure–function relationships, we must now ask what genetic influences can be identified. Insofar as body size and composition are somewhat genetically determined, are there any genetic markers which we can identify to quantify these characteristics? These questions will be taken up in the later chapters. Regardless of the answers to these questions, our data clearly support the hypotheses: (a) There is a strong relationship between structure of the athlete and the specific task (event) in which he excels; and (b) Clear physical prototypes exist for optimal performance at the Olympic Games level.

APPENDIX A: Instructions for Calculating the Heath–Carter Anthropometric Somatotype

Using the Heath–Carter Somatotype Rating Form

The procedure for obtaining the somatotype rating from anthropometric data is described below.

1. Record pertinent identification data in the top section of the rating form. Give the name of the measurer.

First Component (Endomorphy) Rating (Steps 2–5)

2. Record the measurements from each of the four skinfolds.
3. Sum the triceps, subscapular, and suprailiac skinfolds and record in the box opposite TOTAL SKINFOLDS.
4. Circle the closest value in the TOTAL SKINFOLDS scale to the right. (Note: The scale reads vertically from low to high in columns, and horizontally left to right in rows. The rows, "lower limit" and "upper limit," are to provide exact boundaries for each column, and these values should only be circled when the TOTAL SKINFOLDS are within a few millimeters of the limit. In most cases, the value in the row "mid-point" is circled.)
5. Circle the value in the row FIRST COMPONENT which is directly under the column circled in number four above.

Second Component (Mesomorphy) Rating (Steps 6–12)

6. Record the height (inches) and the humerus and femur diameters in the boxes. Before recording the biceps and calf girths in their respective boxes, the corrections for skinfolds must be made. To do this, subtract the triceps skinfold (convert to centimeters first by dividing by 10) from the biceps girth, and subtract the calf skinfold (convert to centimeters) from the calf girth.
7. Mark the point of the subject's height on the height scale which is directly to the right. (Note: Regard the height row as a continuous scale.)
8. For each bone diameter and girth circle the figure in the proper row which is nearest the measurement. (Note: If the measurement falls exactly mid-way between two values, circle the lower value. Because the largest girths and diameters have been recorded the conservative procedure is used.)
9. Now, deal only with columns, not numerical values. Find the column, or space between the columns, that is the average of the column

deviations for the diameters and girths only (not height). To do this:

a. Consider as the zero column the left-most column containing a circled figure.
b. From this zero column, add the total number of columns you must travel horizontally to reach each of the other three circled numbers.
c. Divide this total by four.
d. Take the number obtained by this division and, starting at the zero column, count this number of columns to the right and place a mark (e.g. asterisk) at that point (whether the point be in the middle of a column or a fraction of the way between one column and the next).

10. Still considering columns only, count horizontally the number of columns you must travel from the asterisk to the marked height (or vice versa).

11. From the number 4 in the row marked SECOND COMPONENT move this number of columns to the right or left, depending upon the direction of the asterisk from the height marker. [If the asterisk is to the right of the height marker, move that number of columns to the right of number 4, and if the asterisk is to the left, move left. Because the columns in this row are in half-unit increments, number of columns and half-unit increments (or decrements) are equivalent.]

12. Circle the closest SECOND COMPONENT value determined in 11 above. (If the point is exactly mid-way between two rating points, circle the value closest to the 4 on the scale. This regression toward the 4 is the conservative approach, and is less likely to produce spuriously extreme ratings.)

Third Component (Ectomorphy) Rating (Steps 13–16)

13. Record the weight (pounds).

14. Refer to the nomograph to find the height–weight ratio (HWR, or height/$\sqrt[3]{\text{weight}}$). Record the HWR in the box.

15. Circle the closest value in the HWR scale. (See Note in number 4 earlier.)

16. Locate the THIRD COMPONENT value below the column of the circled HWR and circle it.

Heath–Carter Anthropometric Somatotype

17. Record the circled values obtained above in the appropriate columns in the row Anthropometric Somatotype.

18. The person calculating the rating should sign to the right.

APPENDIX B: Anthropometric Instruments and Measurements

Anthropometric Instruments

Anthropometer (Martin) (A). For the measuring of heights and lengths respectively. Of the four segments, of 50 centimeters each, into which the instrument may be divided, the upper one has a double scale in centimeters and millimeters and was used as a large sliding caliper in which the straight crossbars (B), used in certain measurements, were substituted by curved branches (C) necessary for others.

Small sliding caliper (Martin) (D). Twenty-five centimeters long, was used for smaller dimensions. Following the suggestion of Parnell (1958), a special adaptation was made to fit on to the pointed crossbars in order to increase their lengths and widths and inner surfaces, thus facilitating the measuring of bicondylar widths of humerus and femur.

Measuring tape (E). Narrow and flexible, made out of steel, used to measure circumferences. This kind of tape was preferred to the ones made of fiber in order to avoid its stretching out while being used.

Skinfold caliper gauge (Harpenden) (F). With a constant pressure of 10 gram/millimeter^2 exerted by the surface of its jaws at all openings. All skinfold measurements were taken with this instrument.

Detecto-medic beam-balance scale (G). In kilograms and $\frac{1}{10}$ kilogram for the determination of body weight.

Wall stadiometer (H). Made out of wood and provided with a steel scale graded in centimeters and millimeters for obtaining stature in maximum stretch. This equipment was almost identical to that used by Tanner (1964).

Sitting height bench (I). A rectangular wooden bench similar to that used by Tanner (1964).

Anthropometric Measurements

The majority of measurements were taken with the individual standing erect and straight on a special wooden board completely leveled and with a smooth surface. Only the biepicondylar width of femur and sitting height were taken in a sitting position.

In order to facilitate obtaining certain measurements, the necessary anatomical landmarks: suprasternal, xiphoid, projection of the fourth condro-sternal articulation over the mid line of the body, hypochondral, epigastric, and tibial, were marked by means of a dermographic pencil.

All heights were taken by putting the tip of the left index finger just below the landmark while the right hand lowered the movable crossbar of the anthropometer and placed its point into the correct position on the left index finger.

All measurements were taken to the nearest millimeter.

In those cases where the difference between distances of two somatometric points from the floor were used to obtain the length of a certain segment of the body, as in the case of thoracic and abdominal heights or the lengths of the segments of upper and lower extremities, special care was taken in securing the measurements involved consecutively one right after the other, avoiding any change in the position taken by the subject under observation.

The measurements taken were:

1. *Stature, maximum stretch* (H)[1] was taken with the individual standing straight against the upright wall, touching it with heels, buttocks, and back. The head was oriented in the Frankfurt plane (the upper border of the ear opening and the lower border of the eye socket on a horizontal line), the heels were together and the anterior part of the feet slightly separated from each other. While one person stretched the subject upwards by a gentle traction on the mastoid region, taking care that the heels were kept to the ground, the second one lowered the counterweighted measuring board until it touched the head.

2. *Stature, free-standing* (A). With the subject standing straight but completely free, the head in the Frankfurt Plane, heels together and toes outwards, the anthropometer was placed at the back of the subject and the sliding crossbar was lowered slowly with the right hand until it touched the highest point on the head. With the thumb and the fourth finger of the left hand the observer ascertained that the crossbar touched the scalp in the mid-sagittal plane.

3. *Suprasternal height* (A). The distance of the lowest point of the suprasternal notch from the standing surface.

4. *Xiphoid height* (A). The distance of the point where the base of the xiphoid process is in contact with the sternum, to the standing surface.

5. *Hypochondral height* (A). The distance of the mid-point on the sagittal plane between the xiphoid point and the epigastric point (projection of the lower border of the tenth rib over the mid-sagittal plane) from the standing surface.

6. *Symphyseal height* (A). The distance between the midpoint of the upper border of the pubic crests and the standing surface.

7–8. *Tibial height*, left and right (A). The distance between the supe-

[1] The letter following the measurement stands for the type of instrument used. Each is described in the section titled Anthropometric Instruments.

rior border of the medial condyle of the tibia, corresponding to the interarticular line of the knee joint.

9–10. *Sphyrion height* (foot height), left and right (D). The distance between the inferior point of the medial malleolus and the standing surface. In securing this measurement, the free end of the caliper without the fixed crossbar was set on the standing surface and the movable bar pushed up against the point indicated.

11–12. *Acromial height,* left and right (A). The subject stood with arms straight and tightly placed against the body and the palms of the hands turned inwards. The highest point on the lateral border of the acromion process was determined from the standing surface.

13–14. *Radial height,* left and right (A). The distance between the interarticular line of the elbow, just above the head of the radius, and the standing surface.

15–16. *Styloid height,* left and right (A). The distance of the inferior point of the styloid process of the radius (that is, a point on the wrist close to the base of the thumb) from the standing surface.

17–18. *Dactylion height,* left and right (A). The distance of the lowest point of the middle finger from the standing surface. Special care was taken to keep the hand straight and to avoid its following the curve of the thigh.

19. *Sitting height* (I). The subject sat erect on a table with his feet unsupported. The right angle formed at the knee almost but not completely touched the edge of the table. The hands rested on the thighs. The head was oriented in the Frankfurt plane and the distance between the highest point on the head and the sitting surface was taken, while gentle upward traction was exerted in the mastoid region.

20. *Biacromial breadth* (B). With the individual standing in a relaxed position, the distance between the most lateral points of the acromion processes were measured by the observer who stood behind the subject.

21. *Transverse breadth of the thorax* (B). The distance at the most lateral points of the ribs at the level of the fourth costosternal articulation. The subject was asked to slightly lift his arms while the instrument was placed into position and to lower them again for the reading of the distance. The caliper was held straight in front of the subject and the measurement was taken during the pause between normal inspiration and expiration. This measurement as well as the following ones were taken with the subject standing erect.

22. *Transverse breadth of the thorax* (C). In order to determine the width of the thorax without the influence exerted on the measurement by a strong development of the serratus anterior and latissimus dorsi muscles, the sliding caliper was used with curved crossbars. Its points

were placed laterally on the ribs corresponding to the level of the fourth costosternal articulation. The reading was made after the subject had lowered his arms as far as possible, and during the pause between normal inspiration and expiration.

23. *Hypochondral breadth* (B). With the arms of the subject hanging down loosely, the caliper was held horizontal in front and its crossbars were placed laterally on the level corresponding to the same level at which the hypochondral depth was taken.

24. *Biiliocristal breadth* (B). The horizontal distance between the most lateral points on the superior border of the iliac crest. The observer stood in front of the subject and placed the crossbars of the caliper with both hands on the points sought and considerable presure was exerted on the instrument in order to reduce to a minimum the influence of subcutaneous tissue at this level. The subject stood with his heels together.

25. *Anteroposterior depth of the thorax* (C). The observer stood on the right side of the subject, placing the points of the fixed left crossbar of the caliper with his finger tips against the spine and moving the right one in until it touched a point on the midline of the sternum at the level of the center of the fourth costosternal articulation. This measurement was taken perpendicular to the axis of the thorax and during the pause between normal inspration and expiration.

26. *Hypochondral depth* (C). With the observer in the same position as in the foregoing measurement, the left fixed point of the caliper moved until it touched (without exerting pressure) the hypochondral point marked beforehand. Again, this measurement was taken perpendicular to the axis of the thorax and during the respiratory pause.

27–28. *Biepicondylar breadth of the humerus,* left and right (D). The width between the medial and lateral epicondyles of the humerus, with the arm horizontal and forming a right angle with the forearm. The caliper was applied at an angle approximately bisecting the angle of the elbow. Firm pressure was placed on the crossbars in order to compress the subcutaneous tissue.

29–30. *Biepicondylar breadth of the femur,* left and right (D). With the individual seated and the knee bent at a right angle, the greatest distance between the lateral and medial epicondyles of the femur was determined—with firm pressure on the crossbars.

31–32. *Foot length,* left and right (B). With the subject standing, the distance between the most posterior point of the heel and the tip of the longest toe (usually the first one) was measured. The fixed crossbar of the caliper was applied to the heel and the movable one pushed in.

33–34. *Chest girth at maximum inspiration and expiration* (E). With the arms of the subject slightly lifted, the tape was placed with both hands just below the lower angle of the scapula and then extended towards the front where it coincided in men with the level of the nipples and in women with the level of the fourth costosternal articulation. The subject, standing erect and with his arms hanging loosely at his sides, was then asked to take as deep a breath as possible, and once the maximum inspiration had been reached, the reading was made from the front. The subject was then asked to breathe out as much air as he could, and the reading was again made in front of the subject. In these measurements as well as in the following circumferences, the tape was always in light contact with the skin.

35–36. *Upper arm girth, extended and relaxed,* left and right (E). With the arm extended and hanging loosely at the size of the subject, the tape was placed around it halfway between the acromion process and the elbow joint.

37–38. *Upper arm girth, flexed and tense,* left and right (E). With the arm raised, the elbow flexed and the hand clenched, the elbow flexors were contracted maximally. The measurement was made at the greatest girth of the arm.

39–40. *Minimum girth of the forearm,* left and right (E). The girth was taken with the tape a little higher than the wrist at the level where the lowest reading was obtained on the forearm.

41–42. *Thigh girth,* left and right (E). During this as well as during the following measurement, the subject was standing with his feet slightly apart and with his weight equally distributed on both legs. Perpendicular to the axis of the thigh, the tape was placed around it, with its upper border just below the gluteal fold.

43–44. *Calf girth,* left and right (E). With the subject in the same position as in 41–42, the tape was placed around the calf and the maximum circumference measured.

45–46. *Leg girth, minimum,* left and right (E). The tape was placed above the ankle at the level where the minimum circumference was obtained.

47. *Subscapular skinfold* (F). For this as well as for the following measurements, a fold of skin and subcutaneous tissue was picked up firmly between thumb and forefinger of the left hand and pulled away from the underlying muscle. The plates on the branches of the caliper were then applied a little below the fingers of the left hand and allowed to exert their full pressure before reading the thickness of the fold. All skinfolds were taken on the right side of the body with the subject standing relaxed.

The subscapular skinfold was picked up just beneath the inferior angle of the scapula in a direction which was obliquely downwards and outwards.

48. *Triceps skinfold* (F). With the subject's arm hanging loosely, a fold was picked up at the back of the arm at a level halfway on a line connecting the acromion and the olecranon processes.

49. *Biceps skinfold* (F). At the same level as in the previous measurement a fold was picked up on the anterior of the arm. In this case as well as in the triceps skinfold, the position of the fold was vertical.

50. *Suprailiac skinfold* (F). The fold was picked up three to five centimeters above the anterior superior iliac spine on a diagonal line going downwards and inwards.

51. *Calf lateral skinfold* (F). For this measurement as well as for the following one, the subject placed his right leg in a relaxed position on a chair. At the level of the maximum girth of the calf, a vertical skinfold was picked upon the lateral side of the leg.

52. *Calf medial skinfold* (F). With the individual in the same position as in 51, a vertical fold was picked up at the same level on the inner side of the calf.

53. *Body weight* (G). It was recorded to the nearest $\frac{1}{10}$ kilogram taking care that the subject stood in the middle of the scale platform.

Derived Measurements

The following distances concerning the length of certain segments of the body or differences between maximum and minimum girths were obtained by subtracting one measurement from another. The numbers in parenthesis after the named measurement refer to the measurements numbered and described above from which the differences have been derived.

54. *Sternal length* (3–4).
55. *Total abdominal length* (4–6).
56. *Trunk length* (3–6).
57–58. *Upper arm length,* left (11–13) and right (12–14).
59–60. *Lower arm length,* left (13–15) and right (14–16).
61–62. *Hand length,* left (15–17) and right (16–18).
63–64. *Total arm length,* left (11–15) and right (12–16).
65–66. *Total length of upper extremity,* left (11–17) and right (12–18).
67–68. *Upper leg length,* left (6–7) and right (6–8).
69–70. *Lower leg length,* left (7–9) and right (8–10).
71–72. *Total leg length,* left (6–9) and right (6–10).
73. *Difference between maximum and minimum chest girth* (33–34).

74–75. *Difference between maximum and minimum arm girths,* left (37–35) and right (38–36).

Although a large number of measurements were taken it was not possible to analyze all of them in the present book. Selected measurements are analyzed here and others will be utilized in later studies.

Bibliography

Carter, J. E. L. (1970). The somatotypes of athletes—a review. *Human Biology* **42**, 535–569.

Correnti, V., and Zauli, B. (1964). "Olimpionici 1960." Marves, Roma, Italy.

Hirata, K. (1966). Physique and age of Tokyo olympic champions. *J. Sports Med. and Phys. Fitness* **6**, 207–222.

Jokl, E. (1964). Essay on medical sociology of sports. *In* "Medical Sociology and Cultural Anthropology of Sport and Physical Education." Charles C Thomas, Springfield, Illinois.

Martin, R., and Saller, K. (1959). *Lehrbuch der Anthropologie.* II. Fischer, Stuttgart.

Parnell, R. W. (1958). "Behaviour and Physique." Edward Arnold Ltd., London.

Tanner, J. M. (1964). "The Physique of the Olympic Athlete." George Allen and Unwin, London.

TABLE 3.1. DESCRIPTIVE ANALYSIS FOR TRACKMEN (RUNNERS, HURDLERS, AND WALKERS)

Variable	Units	N	Mean	S.D.	Max.	Min.
Age	yr.	246	24.3	4.3	42.0	16.0
Weight	kg.	244	65.1	7.0	89.7	48.7
Height, Max. stretch	cm.	244	175.0	6.5	196.0	155.8
Trunk length	cm.	243	52.0	2.5	65.2	45.3
Biacromial breadth	cm.	242	39.9	1.8	45.1	35.5
Biiliocristal breadth	cm.	243	27.2	1.6	31.2	23.4
Upper plus lower arm length, right	cm.	243	59.6	3.0	68.0	52.3
Total leg length,right	cm.	240	82.5	4.7	93.2	69.8
Sum 3 skinfolds	mm.	244	17.6	3.3	39.5	11.2
Height-Weight ratio		244	13.17	.38	14.13	12.05
Endomorphy		243	1.6	.43	4.0	1.0
Mesomorphy		243	4.6	.86	8.0	2.0
Ectomorphy		243	3.3	.90	5.5	1.0

TABLE 3.2. DESCRIPTIVE ANALYSIS FOR SPRINTERS (100 m, 200 m,4 x 100 m, 110 m HURDLES)

Variable	Units	N	Mean	S.D.	Max.	Min.
Age	yr.	81	23.9	4.2	39.0	17.0
Weight	kg.	79	68.4	6.6	89.7	54.7
Height, Max. stretch	cm.	79	175.4	6.2	188.7	163.0
Trunk length	cm.	79	51.7	2.2	55.8	45.5
Biacromial breadth	cm.	79	40.3	1.8	45.1	35.5
Biiliocristal breadth	cm.	79	26.8	1.4	29.9	23.5
Upper plus lower arm length, right	cm.	79	60.1	2.8	66.5	54.4
Total leg length,right	cm.	78	83.4	4.5	93.2	72.4
Sum 3 skinfolds	mm.	79	18.9	3.7	39.5	11.7
Height-Weight ratio		79	12.98	.39	14.13	12.05
Endomorphy		78	1.7	.48	4.0	1.0
Mesomorphy		78	5.0	.94	8.0	3.0
Ectomorphy		78	2.8	.91	5.5	1.0

TABLE 3.3. DESCRIPTIVE ANALYSIS FOR 400 m RUNNERS (400 m, 4 x 400 m, 400 m HURDLES)

Variable	Units	N	Mean	S.D.	Max.	Min.
Age	yr.	49	23.4	3.5	30.0	16.0
Weight	kg.	49	67.9	6.3	86.8	56.0
Height, Max. stretch	cm.	49	178.0	5.8	196.0	168.4
Trunk length	cm.	49	52.5	2.2	57.8	47.9
Biacromial breadth	cm.	48	40.4	1.6	43.3	36.0
Biiliocristal breadth	cm.	49	27.2	1.7	31.0	23.5
Upper plus lower arm length, right	cm.	49	60.8	2.5	66.1	55.6
Total leg length,right	cm.	48	84.5	3.7	93.2	78.1
Sum 3 skinfolds	mm.	49	17.0	2.8	25.5	13.1
Height-Weight ratio		49	13.19	.33	13.95	12.50
Endomorphy		49	1.5	.38	2.5	1.0
Mesomorphy		49	4.5	.76	6.0	3.0
Ectomorphy		49	3.3	.81	5.0	1.5

TABLE 3.4. DESCRIPTIVE ANALYSIS FOR MIDDLE DISTANCE RUNNERS (800 m, 1,500 m)

Variable	Units	N	Mean	S.D.	Max.	Min.
Age	yr.	41	22.9	3.0	30.0	17.0
Weight	kg.	41	65.0	5.6	79.4	51.4
Height, Max. stretch	cm.	41	177.3	6.1	191.0	163.8
Trunk length	cm.	41	52.5	3.1	65.2	45.3
Biacromial breadth	cm.	41	40.1	1.7	43.7	37.0
Biiliocristal breadth	cm.	41	27.6	2.0	31.2	24.3
Upper plus lower arm length, right	cm.	41	60.1	3.0	68.0	54.0
Total leg length,right	cm.	41	84.0	4.7	92.2	74.7
Sum 3 skinfolds	mm.	41	16.8	1.9	21.8	12.8
Height-Weight ratio		41	13.33	.30	13.95	12.50
Endomorphy		41	1.5	.25	2.0	1.0
Mesomorphy		41	4.2	.61	5.5	3.0
Ectomorphy		41	3.6	.72	5.0	1.5

TABLE 3.5. DESCRIPTIVE ANALYSIS FOR LONG DISTANCE RUNNERS (3,000 m STEEPLECHASE, 5,000 m, 10,000 m)

Variable	Units	N	Mean	S.D.	Max.	Min.
Age	yr.	34	25.3	4.5	42.0	20.0
Weight	kg.	34	59.8	5.4	71.7	48.7
Height, Max. stretch	cm.	34	171.9	5.5	186.0	155.8
Trunk length	cm.	34	51.2	2.4	55.9	46.1
Biacromial breadth	cm.	34	39.1	1.5	42.7	35.9
Biiliocristal breadth	cm.	34	27.2	1.4	29.7	23.4
Upper plus lower arm length, right	cm.	34	58.4	2.5	62.6	52.3
Total leg length,right	cm.	33	80.3	3.3	87.9	72.2
Sum 3 skinfolds	mm.	34	16.6	3.9	27.5	11.2
Height-Weight ratio		34	13.31	.36	13.95	12.50
Endomorphy		34	1.4	.50	3.0	1.0
Mesomorphy		34	4.1	.74	5.5	2.5
Ectomorphy		34	3.6	.89	5.0	1.5

TABLE 3.6. DESCRIPTIVE ANALYSIS FOR MARATHON RUNNERS

Variable	Units	N	Mean	S.D.	Max.	Min.
Age	yr.	20	26.4	4.8	36.0	20.0
Weight	kg.	20	56.6	3.7	66.1	51.7
Height, Max. stretch	cm.	20	168.7	6.3	176.6	159.0
Trunk length	cm.	20	50.6	1.8	54.5	46.9
Biacromial breadth	cm.	20	38.5	1.5	41.8	36.2
Biiliocristal breadth	cm.	20	26.9	1.0	29.1	25.0
Upper plus lower arm length, right	cm.	20	56.4	2.3	61.3	53.4
Total leg length, right	cm.	20	77.5	4.2	82.7	69.8
Sum 3 skinfolds	mm.	20	16.7	3.8	24.9	11.9
Height-Weight ratio		20	13.27	.40	14.10	12.76
Endomorphy		20	1.4	.49	2.5	1.0
Mesomorphy		20	4.3	.81	5.5	2.0
Ectomorphy		20	3.5	.94	5.5	2.5

TABLE 3.7. DESCRIPTIVE ANALYSIS FOR WALKERS (20 km WALK, 50 km WALK)

Variable	Units	N	Mean	S.D.	Max.	Min.
Age	yr.	21	27.1	5.6	37.0	20.0
Weight	kg.	21	62.9	5.1	73.8	51.3
Height, Max. stretch	cm.	21	173.2	5.6	184.7	165.5
Trunk length	cm.	20	53.4	2.2	57.3	49.8
Biacromial breadth	cm.	20	39.6	1.5	42.6	36.5
Biiliocristal breadth	cm.	20	27.9	1.0	30.2	26.5
Upper plus lower arm length, right	cm.	20	58.6	2.8	64.8	53.1
Total leg length, right	cm.	20	79.7	4.2	89.8	73.7
Sum 3 skinfolds	mm.	21	18.0	1.8	21.8	15.5
Height-Weight ratio		21	13.22	.27	13.85	12.81
Endomorphy		21	1.6	.22	2.0	1.5
Mesomorphy		21	4.7	.66	6.0	3.5
Ectomorphy		21	3.4	.66	5.0	2.5

TABLE 3.8. FREQUENCIES OF SOMATOTYPES ACCORDING TO COMPONENT DOMINANCE

Somatotype Group	Track Short dist.		Track Long dist.		Field Jumpers		Field Throwers		Track and Field Decathlon		Track and Field Total	
	N	%	N	%	N	%	N	%	N	%	N	%
Endomorphy												
Endo-Meso	2	1					14	61			16	5
Mesomorphy	50	36	23	21	11	29	9	39	7	88	100	32
Ecto-Meso	77	56	75	69	23	61			1	13	176	56
Ectomorphy	3	2	4	4	1	3					8	3
Central	5	4	6	6	3	8					14	4
Totals	137	99	108	100	38	101	23	100	8	101	314	100

TABLE 3.9. ANALYSIS OF DIFFERENCES BETWEEN GROUPS - TRACK (MEN)

Variable	F-ratio	Between group differences					
Height	10.52**	Marathon	Long Distance	Walk	Sprint	Middle Distance	400 m
Weight	21.54**	Marathon	Long Distance	Walk	Middle Distance	400 m	Sprint
Endomorphy	4.33**	Long Distance	Marathon	Middle Distance	400 m	Walk	Sprint
Mesomorphy	9.54**	Long Distance	Middle Distance	Marathon	400 m	Walk	Sprint
Ectomorphy	7.59**	Sprint	400 m	Walk	Marathon	Long Distance	Middle Distance
Biacromiale	4.88**	Marathon	Long Distance	Walk	Middle Distance	400 m	Sprint
Biiliocristale	2.89*	Sprint	Marathon	Long Distance	400 m	Middle Distance	Walk
Arm length	10.24**	Marathon	Long Distance	Walk	Sprint	Middle Distance	400 m
Leg length	12.85**	Marathon	Walk	Long Distance	Sprint	Middle Distance	400 m
Trunk length	4.72**	Marathon	Long Distance	Sprint	Middle Distance	400 m	Walk

* $F_{.05}(5,238) = 2.25$ ** $F_{.01}(5,238) = 3.10$

TABLE 3.10. DESCRIPTIVE ANALYSIS FOR FIELD (MEN)

Variable	Units	N	Mean	S.D.	Max.	Min.
Age	yr.	71	25.2	4.5	43.0	17.0
Weight	kg.	69	80.4	14.8	126.2	57.7
Height, Max. stretch	cm.	69	182.8	7.1	204.3	169.9
Trunk length	cm.	68	54.5	2.9	64.0	48.7
Biacromial breadth	cm.	68	42.3	2.3	48.1	37.1
Biiliocristal breadth	cm.	68	28.9	2.2	34.2	24.1
Upper plus lower arm length, right	cm.	68	62.3	3.6	70.0	53.8
Total leg length, right	cm.	68	85.9	4.7	99.6	75.9
Sum 3 skinfolds	mm.	69	22.2	10.6	72.7	13.0
Height-Weight ratio		69	12.85	.55	13.80	11.25
Endomorphy		69	2.1	1.10	6.5	1.0
Mesomorphy		69	5.3	1.32	9.0	3.0
Ectomorphy		69	2.6	1.13	5.0	.5

TABLE 3.11. DESCRIPTIVE ANALYSIS FOR JUMPERS (HIGH JUMP, LONG JUMP, TRIPLE JUMP)

Variable	Units	N	Mean	S.D.	Max.	Min.
Age	yr.	31	23.5	3.5	29.0	17.0
Weight	kg.	31	73.2	7.9	89.8	57.7
Height, Max. stretch	cm.	31	182.8	8.2	204.3	170.2
Trunk length	cm.	31	53.6	2.4	59.6	49.7
Biacromial breadth	cm.	31	41.4	2.0	45.1	37.1
Biiliocristal breadth	cm.	31	27.7	2.1	32.5	24.1
Upper plus lower arm length, right	cm.	31	62.4	3.8	70.0	56.1
Total leg length, right	cm.	31	87.2	5.4	99.6	77.3
Sum 3 skinfolds	mm.	31	18.1	3.0	24.0	13.0
Height-Weight ratio		31	13.20	.30	13.80	12.65
Endomorphy		31	1.7	.38	2.5	1.0
Mesomorphy		31	4.4	.76	6.0	3.0
Ectomorphy		31	3.4	.75	5.0	2.0

TABLE 3.12. DESCRIPTIVE ANALYSIS FOR POLE VAULT

Variable	Units	N	Mean	S.D.	Max.	Min.
Age	yr.	6	24.5	3.3	27.0	19.0
Weight	kg.	6	70.6	6.3	79.1	59.6
Height, Max. stretch	cm.	6	179.7	4.9	182.8	169.9
Trunk length	cm.	6	54.0	2.2	56.9	51.5
Biacromial breadth	cm.	6	41.4	1.7	43.7	39.5
Biiliocristal breadth	cm.	6	29.3	1.7	31.9	26.5
Upper plus lower arm length, right	cm.	6	60.6	4.4	65.6	53.8
Total leg length, right	cm.	6	83.2	4.1	87.7	75.9
Sum 3 skinfolds	mm.	6	16.9	2.9	20.7	13.1
Height-Weight ratio		6	13.13	.17	13.29	12.82
Endomorphy		6	1.5	.45	2.0	1.0
Mesomorphy		6	4.8	.41	5.5	4.5
Ectomorphy		6	3.2	.41	3.5	2.5

TABLE 3.13. DESCRIPTIVE ANALYSIS FOR WEIGHT THROWERS (SHOT, DISCUS, HAMMER)

Variable	Units	N	Mean	S.D.	Max.	Min.
Age	yr.	15	27.3	5.3	43.0	21.0
Weight	kg.	14	102.3	12.2	126.2	86.0
Height, Max. stretch	cm.	14	186.1	5.6	195.9	177.4
Trunk length	cm.	14	56.7	3.0	62.7	51.4
Biacromial breadth	cm.	14	44.4	2.4	48.1	40.0
Biiliocristal breadth	cm.	14	30.6	2.2	34.2	26.8
Upper plus lower arm length, right	cm.	14	63.8	3.7	69.6	56.7
Total leg length, right	cm.	14	86.1	4.2	91.5	78.2
Sum 3 skinfolds	mm.	14	35.0	16.9	72.7	15.0
Height-Weight ratio		14	12.01	.42	12.80	11.75
Endomorphy		14	3.5	1.55	6.5	1.5
Mesomorphy		14	7.1	1.00	9.0	5.5
Ectomorphy		14	1.0	.55	2.5	.5

TABLE 3.14. DESCRIPTIVE ANALYSIS FOR JAVELIN THROWERS

Variable	Units	N	Mean	S.D.	Max.	Min.
Age	yr.	8	28.4	5.4	36.0	21.0
Weight	kg.	8	76.7	7.3	84.6	63.4
Height, Max. stretch	cm.	8	179.5	4.7	186.5	174.7
Trunk length	cm.	7	53.8	2.8	56.7	48.7
Biacromial breadth	cm.	7	42.0	1.7	43.9	38.7
Biiliocristal breadth	cm.	7	29.7	.5	30.6	29.2
Upper plus lower arm length, right	cm.	7	60.7	2.0	62.9	57.1
Total leg length,right	cm.	7	82.6	3.3	86.6	76.6
Sum 3 skinfolds	mm.	8	21.3	7.9	35.4	14.2
Height-Weight ratio		8	12.77	.33	13.35	12.37
Endomorphy		8	2.0	.92	3.5	1.0
Mesomorphy		8	5.6	.78	7.0	4.5
Ectomorphy		8	2.3	.75	3.5	1.5

TABLE 3.15. DESCRIPTIVE ANALYSIS FOR DECATHLON

Variable	Units	N	Mean	S.D.	Max.	Min.
Age	yr.	9	25.1	3.6	31.0	20.0
Weight	kg.	8	77.5	3.4	82.3	73.3
Height, Max. stretch	cm.	8	181.3	2.2	183.9	177.0
Trunk length	cm.	8	54.2	1.8	56.4	50.3
Biacromial breadth	cm.	8	42.8	1.4	44.5	40.3
Biiliocristal breadth	cm.	8	29.1	1.6	31.3	26.7
Upper plus lower arm length, right	cm.	8	61.0	1.1	62.1	58.6
Total leg length,right	cm.	8	84.6	1.7	87.0	81.3
Sum 3 skinfolds	mm.	8	19.9	2.4	23.3	17.7
Height-Weight ratio		8	12.84	.25	13.23	12.41
Endomorphy		8	1.8	.37	2.5	1.5
Mesomorphy		8	5.6	.83	6.5	4.0
Ectomorphy		8	2.5	.60	3.5	1.5

TABLE 3.16. ANALYSIS OF DIFFERENCES BETWEEN GROUPS - FIELD (MEN)

Variable	F-ratio	Between group differences				
Height	2.34	Javelin	Vault	Decathlon	Jump	Weight
Weight	35.06**	Vault	Jump	Javelin	Decathlon	Weight
(Weight)+	(1.58)	(Vault	Jump	Javelin	Decathlon)	
Endomorphy	11.35**	Vault	Jump	Decathlon	Javelin	Weight
Mesomorphy	25.60**	Jump	Vault	Javelin	Decathlon	Weight
Ectomorphy	34.12**	Weight	Javelin	Decathlon	Vault	Jump
Biacromiale	7.14**	Vault	Jump	Javelin	Decathlon	Weight
Biiliocristale	6.63**	Jump	Decathlon	Vault	Javelin	Weight
Arm length	1.89	Vault	Javelin	Decathlon	Jump	Weight
Leg length	2.36	Javelin	Vault	Decathlon	Weight	Jump
Trunk length	5.13**	Jump	Javelin	Vault	Decathlon	Weight

* $F_{.05}$ (4,63) = 2.52 ** $F_{.01}$ (4,63) = 3.65 + Analysis of variance run without the weight throwers group.

TABLE 3.17. DESCRIPTIVE ANALYSIS FOR TRACK AND FIELD, WOMEN

Variable	Units	N	Mean	S.D.	Max.	Min.
Age	yr.	82	20.8	3.6	29.0	15.0
Weight	kg.	82	59.2	9.7	97.7	45.3
Height, Max. stretch	cm.	82	168.0	7.5	188.5	151.1
Trunk length	cm.	82	49.7	3.1	56.6	42.0
Biacromial breadth	cm.	82	37.2	1.9	42.6	33.4
Biiliocristal breadth	cm.	82	27.2	2.4	35.3	22.9
Upper plus lower arm length, right	cm.	82	56.3	2.9	63.8	48.3
Total leg length,right	cm.	82	79.4	4.8	95.0	68.0
Sum 3 skinfolds	mm.	82	28.3	14.1	91.4	14.1
Height-Weight ratio		82	13.09	.47	14.05	11.31
Endomorphy		82	2.8	1.40	8.0	1.0
Mesomorphy		82	3.8	.96	7.5	2.0
Ectomorphy		82	3.1	1.05	5.5	.5

TABLE 3.18. DESCRIPTIVE ANALYSIS FOR SPRINTERS, WOMEN (100 m, 200 m, 4 x 100 m, 80 m HURDLES)

Variable	Units	N	Mean	S.D.	Max.	Min.
Age	yr.	28	20.7	3.4	29.0	15.0
Weight	kg.	28	56.8	6.6	74.4	45.3
Weight, Max. stretch	cm.	28	165.0	6.3	180.8	151.7
Trunk length	cm.	28	49.0	3.0	55.8	43.4
Biacromial breadth	cm.	28	36.7	1.7	39.7	33.4
Biiliocristal breadth	cm.	28	26.2	1.7	28.4	22.9
Upper plus lower arm length, right	cm.	28	55.3	2.7	59.6	48.3
Total leg length,right	cm.	28	77.8	4.6	85.7	68.0
Sum 3 skinfolds	mm.	28	27.2	8.1	49.0	16.1
Height-Weight ratio		28	13.01	.33	13.65	12.35
Endomorphy		28	2.7	.93	5.0	1.5
Mesomorphy		28	3.9	.66	5.5	2.5
Ectomorphy		28	2.9	.80	4.5	1.5

TABLE 3.19. DESCRIPTIVE ANALYSIS FOR MIDDLE DISTANCE, WOMEN (400 m, 800 m)

Variable	Units	N	Mean	S.D.	Max.	Min.
Age	yr.	18	20.0	2.8	25.0	15.0
Weight	kg.	18	54.3	4.2	60.9	46.9
Height, Max. stretch	cm.	18	166.9	6.0	176.2	151.1
Trunk length	cm.	18	49.2	2.3	53.6	45.3
Biacromial breadth	cm.	18	36.6	1.3	38.5	34.2
Biiliocristal breadth	cm.	18	27.1	1.7	29.5	24.0
Upper plus lower arm length, right	cm.	18	55.9	2.5	61.2	51.1
Total leg length,right	cm.	18	78.8	3.5	87.7	70.8
Sum 3 skinfolds	mm.	18	21.3	5.1	31.6	14.1
Height-Weight ratio		18	13.35	.30	14.00	12.70
Endomorphy		18	2.0	.72	3.5	1.0
Mesomorphy		18	3.3	.64	4.0	2.0
Ectomorphy		18	3.7	.77	5.5	2.0

TABLE 3.20. DESCRIPTIVE ANALYSIS FOR JUMPING, WOMEN (HIGH JUMP, LONG JUMP)

Variable	Units	N	Mean	S.D.	Max.	Min.
Age	yr.	12	21.5	4.1	28.0	15.0
Weight	kg.	12	56.4	4.3	66.4	49.6
Height, Max. stretch	cm.	12	169.4	7.7	188.5	159.7
Trunk length	cm.	12	48.0	2.8	51.4	42.0
Biacromial breadth	cm.	12	37.2	1.8	40.9	34.8
Biiliocristal breadth	cm.	12	25.9	2.2	29.6	23.2
Upper plus lower arm length, right	cm.	12	58.0	2.9	63.8	53.6
Total leg length,right	cm.	12	82.3	5.7	95.0	75.9
Sum 3 skinfolds	mm.	12	22.2	4.5	31.7	15.6
Height-Weight ratio		12	13.39	.35	14.05	12.90
Endomorphy		12	2.2	.54	3.5	1.5
Mesomorphy		12	3.3	.68	4.5	2.0
Ectomorphy		12	3.7	.89	5.5	2.5

TABLE 3.21. DESCRIPTIVE ANALYSIS FOR WEIGHT THROWING, WOMEN (SHOT-PUT, DISCUS)

Variable	Units	N	Mean	S.D.	Max.	Min.
Age	yr.	9	19.9	4.3	28.0	15.0
Weight	kg.	9	73.5	12.6	89.3	52.0
Height, Max. stretch	cm.	9	170.9	8.3	186.5	163.0
Trunk length	cm.	9	51.3	4.1	56.6	44.0
Biacromial breadth	cm.	9	38.3	2.9	42.6	34.4
Biiliocristal breadth	cm.	9	30.0	2.5	32.3	25.2
Upper plus lower arm length, right	cm.	9	56.4	3.7	62.4	52.6
Total leg length,right	cm.	9	79.8	4.9	85.5	70.9
Sum 3 skinfolds	mm.	9	54.8	22.8	91.4	15.6
Height-Weight ratio		9	12.38	.63	13.71	11.31
Endomorphy		9	5.3	1.97	8.0	1.5
Mesomorphy		9	5.2	1.39	7.5	3.0
Ectomorphy		9	1.7	1.14	4.5	.5

TABLE 3.22. DESCRIPTIVE ANALYSIS FOR JAVELIN THROWING, WOMEN

Variable	Units	N	Mean	S.D.	Max.	Min.
Age	yr.	4	20.2	3.4	25.0	17.0
Weight	kg.	4	73.1	16.8	97.7	61.5
Height, Max. stretch	cm.	4	180.1	5.5	186.4	173.5
Trunk length	cm.	4	54.0	.7	54.8	53.1
Biacromial breadth	cm.	4	39.3	2.7	42.3	35.8
Biiliocristal breadth	cm.	4	30.1	3.9	35.3	26.5
Upper plus lower arm length, right	cm.	4	59.6	1.4	61.1	57.7
Total leg length,right	cm.	4	84.6	4.7	90.1	79.2
Sum 3 skinfolds	mm.	4	36.4	18.5	63.0	20.6
Height-Weight ratio		4	13.11	.71	13.95	12.22
Endomorphy		4	3.6	1.70	6.0	2.0
Mesomorphy		4	3.7	1.26	5.0	2.0
Ectomorphy		4	3.1	1.6	5.0	1.0

TABLE 3.23. DESCRIPTIVE ANALYSIS FOR PENTATHLON, WOMEN

Variable	Units	N	Mean	S.D.	Max.	Min.
Age	yr.	11	22.5	4.4	29.0	15.0
Weight	kg.	11	60.0	7.4	72.8	50.9
Height, Max. stretch	cm.	11	169.5	7.4	180.4	154.7
Trunk length	cm.	11	51.4	2.1	54.2	47.0
Biacromial breadth	cm.	11	37.8	1.5	40.3	35.5
Biiliocristal breadth	cm.	11	27.9	2.1	32.6	24.9
Upper plus lower arm length, right	cm.	11	56.3	2.8	60.8	51.0
Total leg length,right	cm.	11	79.0	4.3	85.9	71.5
Sum 3 skinfolds	mm.	11	24.5	6.3	36.6	15.7
Height-Weight ratio		11	13.12	.28	13.51	12.64
Endomorphy		11	2.4	.79	4.0	1.5
Mesomorphy		11	3.6	.73	5.0	2.5
Ectomorphy		11	3.1	.66	4.0	2.0

TABLE 3.24. ANALYSIS OF DIFFERENCES BETWEEN GROUPS - TRACK AND FIELD (WOMEN)

Variable	F-ratio	Between group differences					
Height	4.20**	Sprint	Middle Distance	Jump	Pentathlon	Weight	Javelin
Weight	2.79*	Middle Distance	Jump	Sprint	Pentathlon		
Endomorphy	15.39**	Middle Distance	Jump	Pentathlon	Sprint	Javelin	Weight
Mesomorphy	7.64**	Middle Distance	Jump	Pentathlon	Javelin	Sprint	Weight
Ectomorphy	9.48**	Weight	Sprint	Javelin	Pentathlon	Middle Distance	Jump
Biacromiale	2.65*	Middle Distance	Sprint	Jump	Pentathlon	Weight	Javelin
Biiliocristale	7.57**	Jump	Sprint	Middle Distance	Pentathlon	Weight	Javelin
Arm length	2.83*	Sprint	Middle Distance	Pentathlon	Weight	Jump	Javelin
Leg length	2.81*	Sprint	Middle Distance	Pentathlon	Weight	Jump	Javelin
Trunk length	4.66**	Jump	Sprint	Middle Distance	Weight	Pentathlon	Javelin

* $F_{.05}$ (5,76) = 2.34 ** $F_{.01}$ (5.76) = 3.27

TABLE 3.25. FREQUENCIES OF SOMATOTYPES IN TRACK AND FIELD (WOMEN) ACCORDING TO COMPONENT DOMINANCE

Somatotype Group	Track Sprinters		Track Middle Distance		Field Jumpers		Field Throwers		Track and Field Pentathlon		Track and Field Total	
	N	%	N	%	N	%	N	%	N	%	N	%
Endomorphy							2	15			2	2
Endo-Meso	2	7					8	62			10	12
Mesomorphy	4	14	1	6	1	8			2	18	8	10
Ecto-Meso	6	21	5	28	1	8			2	18	14	17
Ectomorphy	1	4	2	11	2	17	1	8			6	7
Central	15	54	10	56	8	67	2	15	7	64	42	51
Totals	28	100	18	101	12	100	13	100	11	100	82	99

TABLE 3.26. DESCRIPTIVE ANALYSIS FOR SWIMMING (MEN), TOTAL GROUP

Variable	Units	N	Mean	S.D.	Max.	Min.
Age	yr.	67	19.2	2.4	25.0	14.0
Weight	kg.	66	72.1	6.8	88.9	57.6
Height, Max. stretch	cm.	66	179.3	6.2	190.0	164.3
Trunk length	cm.	66	54.4	2.7	62.0	48.7
Biacromial breadth	cm.	66	41.2	2.0	45.0	34.5
Biiliocristal breadth	cm.	66	28.1	1.4	32.0	25.4
Upper plus lower arm length, right	cm.	65	60.7	3.0	69.5	54.5
Total leg length,right	cm.	65	82.1	3.7	88.2	72.4
Sum 3 skinfolds	mm.	66	22.0	5.1	36.3	14.5
Height-Weight ratio		66	13.02	0.30	13.65	12.43
Endomorphy		65	2.1	0.63	4.0	1.0
Mesomorphy		65	5.0	0.80	7.0	4.0
Ectomorphy		65	2.9	0.71	4.5	1.5

TABLE 3.27. DESCRIPTIVE ANALYSIS FOR SWIMMING MEN, FREE STYLE

Variable	Units	N	Mean	S.D.	Max.	Min.
Age	yr.	22	20.0	2.2	25.0	17.0
Weight	kg.	22	74.2	7.1	86.6	62.1
Height, Max. stretch	cm.	22	181.7	5.9	190.0	168.7
Trunk length	cm.	22	55.0	2.4	58.4	50.9
Biacromial breadth	cm.	22	41.7	1.7	45.0	36.3
Biiliocristal breadth	cm.	22	28.6	1.4	31.1	26.8
Upper plus lower arm length, right	cm.	22	62.0	2.6	66.8	57.1
Total leg length,right	cm.	21	83.4	3.9	88.2	75.1
Sum 3 skinfolds	mm.	22	22.9	4.8	33.4	17.6
Height-Weight ratio		22	13.07	.24	13.52	12.73
Endomorphy		21	2.2	.64	3.5	1.5
Mesomorphy		21	4.7	.72	6.0	4.0
Ectomorphy		21	2.9	.58	4.0	2.0

TABLE 3.28. DESCRIPTIVE ANALYSIS FOR SWIMMING (MEN), BREASTSTROKE

Variable	Units	N	Mean	S.D.	Max.	Min.
Age	yr.	12	18.6	2.4	24.0	16.0
Weight	kg.	12	69.0	6.2	80.2	61.4
Height, Max. stretch	cm.	12	175.4	6.3	185.3	164.3
Trunk length	cm.	12	53.2	2.0	55.7	50.2
Briacromial breadth	cm.	12	40.4	2.5	43.6	34.5
Biiliocristal breadth	cm.	12	26.8	0.9	28.5	25.4
Upper plus lower arm length, right	cm.	12	59.4	3.5	65.5	54.5
Total leg length,right	cm.	12	79.8	4.2	88.4	72.4
Sum 3 skinfolds	mm.	12	22.1	4.8	35.6	15.9
Height-Weight ratio		12	12.93	0.34	13.65	12.55
Endomorphy		12	2.2	0.49	3.5	1.5
Mesomorphy		12	5.3	0.81	6.0	4.0
Ectomorphy		12	2.8	0.75	4.5	2.0

TABLE 3.29. DESCRIPTIVE ANALYSIS FOR SWIMMING MEN, BACK STROKE

Variable	Units	N	Mean	S.D.	Max.	Min.
Age	yr.	11	17.7	1.7	20.0	14.0
Weight	kg.	11	70.3	5.8	81.1	57.6
Height, Max. stretch	cm.	11	180.3	5.3	188.0	173.2
Trunk length	cm.	11	54.6	3.5	60.0	48.7
Biacromial breadth	cm.	11	41.1	1.9	44.3	38.6
Biiliocristal breadth	cm.	11	27.7	1.1	29.7	25.8
Upper plus lower arm length, right	cm.	10	61.7	3.4	69.5	58.1
Total leg length,right	cm.	11	83.6	3.6	89.5	79.0
Sum 3 skinfolds	mm.	11	21.5	7.3	36.3	14.5
Height-Weight ratio		11	13.20	.33	13.65	12.76
Endomorphy		11	2.1	.92	4.0	1.0
Mesomorphy		11	4.6	.62	6.0	4.0
Ectomorphy		11	3.4	.80	4.5	2.5

TABLE 3.30. DESCRIPTIVE ANALYSIS FOR SWIMMING MEN, BUTTERFLY

Variable	Units	N	Mean	S.D.	Max.	Min.
Age	yr.	11	20.1	2.5	24.0	16.0
Weight	kg.	10	72.2	7.1	88.9	64.0
Height, Max. stretch	cm.	10	178.2	5.9	188.9	171.2
Trunk length	cm.	10	54.7	3.4	62.0	49.6
Biacromial breadth	cm.	10	40.7	1.9	44.1	38.5
Biiliocristal breadth	cm.	10	28.5	1.7	32.0	26.4
Upper plus lower arm length, right	cm.	10	59.5	2.7	63.7	56.6
Total leg length,right	cm.	10	81.1	2.9	85.9	76.6
Sum 3 skinfolds	mm.	10	21.4	5.2	34.8	16.8
Height-Weight ratio		10	12.95	.36	13.52	12.43
Endomorphy		10	2.0	.62	3.5	1.5
Mesomorphy		10	5.2	.95	7.0	4.0
Ectomorphy		10	2.6	.81	4.0	1.5

TABLE 3.31. DESCRIPTIVE ANALYSIS FOR SWIMMING, MEN, MEDLEY

Variable	Units	N	Mean	S.D.	Max.	Min.
Age	yr.	12	19.0	2.4	24.0	15.0
Weight	kg.	12	73.0	6.3	80.6	61.0
Height, Max. stretch	cm.	12	178.5	5.8	187.9	168.5
Trunk length	cm.	12	53.9	2.7	59.4	50.7
Briacromial breadth	cm.	12	41.8	1.7	44.2	38.8
Biiliocristal breadth	cm.	12	28.3	1.1	29.8	26.8
Upper plus lower arm length, right	cm.	12	60.0	1.5	61.7	56.6
Total leg length,right	cm.	12	82.1	2.8	86.5	78.0
Sum 3 skinfolds	mm.	12	20.3	3.8	28.4	15.1
Height-Weight ratio		12	12.91	.20	13.21	12.66
Endomorphy		12	1.9	.47	3.0	1.5
Mesomorphy		12	5.3	.69	6.0	4.5
Ectomorphy		12	2.7	.58	3.5	2.0

TABLE 3.32. FREQUENCIES OF SOMATOTYPES IN SWIMMING (MEN) ACCORDING TO COMPONENT DOMINANCE

Somatotype Group	Freestyle		Breaststroke		Backstroke		Butterfly		Medley		Total	
	N	%	N	%	N	%	N	%	N	%	N	%
Endomorphy												
Endo-Meso							1	10			1	2
Mesomorphy	10	48	8	67	2	20	5	50	9	75	34	52
Ecto-Meso	5	24	2	17	5	50	3	30	3	25	18	28
Ectomorphy												
Central	6	29	2	17	3	30	1	10			12	18
Totals	21	101	12	101	10	100	10	100	12	100	65	100

TABLE 3.33. ANALYSIS OF DIFFERENCES BETWEEN GROUPS - SWIMMING (MEN)

Variable	F-ratio	Between group differences				
Height	2.52*	Breast	Butterfly	Medley	Back	Free
Weight	1.44	Breast	Back	Butterfly	Medley	Free
Endomorphy	0.53	Medley	Butterfly	Breast	Back	Free
Mesomorphy	2.82*	Back	Free	Butterfly	Medley	Breast
Ectomorphy	2.33	Medley	Breast	Butterfly	Free	Back
Biacromiale	1.31	Breast	Butterfly	Back	Free	Medley
Biiliocristale	4.30**	Breast	Back	Medley	Butterfly	Free
Arm length	1.20	Back	Breast	Butterfly	Medley	Free
Leg length	2.31	Breast	Butterfly	Medley	Free	Back
Trunk length	0.90	Breast	Medley	Butterfly	Free	Back

* $F_{.05}$ (4.61) = 2.52 ** $F_{.01}$ (4.61) = 3.65

TABLE 3.34. DESCRIPTIVE ANALYSIS FOR SWIMMING WOMEN, TOTAL GROUP

Variable	Units	N	Mean	S.D.	Max.	Min.
Age	yr.	32	16.3	2.9	23.0	12.0
Weight	kg.	28	56.9	9.1	81.0	40.0
Height, Max. stretch	cm.	28	164.4	7.1	178.0	152.0
Trunk length	cm.	28	49.5	2.7	55.1	44.6
Biacromial breadth	cm.	28	37.1	1.7	39.5	33.3
Biiliocristal breadth	cm.	29	27.1	2.0	31.0	22.4
Upper plus lower arm length, right	cm.	29	55.1	2.9	62.3	48.6
Total leg length, right	cm.	29	76.1	4.4	84.3	68.4
Sum 3 skinfolds	mm.	28	34.2	17.7	110.8	18.2
Height-Weight ratio		28	13.03	.37	14.00	12.32
Endomorphy		28	3.4	1.50	9.0	1.5
Mesomorphy		28	4.0	.71	5.5	2.5
Ectomorphy		28	3.0	.95	5.5	1.0

TABLE 3.35. DESCRIPTIVE ANALYSIS FOR SWIMMING WOMEN, FREESTYLE

Variable	Units	N	Mean	S.D.	Max.	Min.
Age	yr.	8	15.2	2.9	20.0	12.0
Weight	kg.	7	53.4	10.2	68.8	40.0
Height, Max. stretch	cm.	7	165.9	8.7	178.0	156.2
Trunk length	cm.	7	49.1	3.0	52.6	44.6
Biacromial breadth	cm.	7	36.8	2.4	39.5	33.3
Biiliocristal breadth	cm.	7	26.1	2.8	30.8	22.4
Upper plus lower arm length, right	cm.	7	55.3	3.6	62.3	50.9
Total leg length, right	cm.	7	77.7	4.0	83.8	73.4
Sum 3 skinfolds	mm.	7	31.2	10.6	51.5	20.4
Height-Weight ratio		7	13.39	.33	14.00	13.02
Endomorphy		7	3.1	1.06	5.0	2.0
Mesomorphy		7	3.6	.45	4.0	3.0
Ectomorphy		7	3.9	.90	5.5	3.0

TABLE 3.36. DESCRIPTIVE ANALYSIS FOR SWIMMING WOMEN, BREASTSTROKE

Variable	Units	N	Mean	S.D.	Max.	Min.
Age	yr.	9	16.8	4.2	23.0	12.0
Weight	kg.	8	53.9	7.3	62.8	44.3
Height, Max. stretch	cm.	8	160.9	7.2	169.3	152.0
Trunk length	cm.	8	49.4	3.6	55.1	45.0
Biacromial breadth	cm.	8	36.8	1.8	39.3	34.4
Biiliocristal breadth	cm.	9	26.9	1.6	29.1	24.3
Upper plus lower arm length, right	cm.	9	55.4	1.7	58.3	53.4
Total leg length,right	cm.	9	73.9	3.6	79.4	68.4
Sum 3 skinfolds	mm.	8	34.9	31.1	110.8	18.2
Height-Weight ratio		8	12.92	.21	13.30	12.61
Endomorphy		8	3.2	2.43	9.0	1.5
Mesomorphy		8	4.2	.84	5.5	2.5
Ectomorphy		8	2.7	.46	3.5	2.0

TABLE 3.37. DESCRIPTIVE ANALYSIS FOR SWIMMING WOMEN, BACK STROKE

Variable	Units	N	Mean	S.D.	Max.	Min.
Age	yr.	4	17.5	1.7	19.0	16.0
Weight	kg.	3	57.2	3.1	60.8	55.2
Height, Max. stretch	cm.	3	164.6	6.4	170.1	157.6
Trunk length	cm.	3	49.7	1.4	51.0	48.2
Biacromial breadth	cm.	3	36.2	1.3	37.4	34.8
Biiliocristal breadth	cm.	3	27.6	.8	28.5	27.1
Upper plus lower arm length, right	cm.	3	54.0	2.6	57.0	52.1
Total leg length,right	cm.	3	75.1	2.5	77.7	72.6
Sum 3 skinfolds	mm.	3	32.1	11.9	43.8	20.0
Height-Weight ratio		3	12.93	.34	13.17	12.54
Endomorphy		3	3.3	1.26	4.5	2.0
Mesomorphy		3	4.5	.50	5.0	4.0
Ectomorphy		3	2.8	.76	3.5	2.0

TABLE 3.38. DESCRIPTIVE ANALYSIS FOR SWIMMING WOMEN, MEDLEY

Variable	Units	N	Mean	S.D.	Max.	Min.
Age	yr.	11	16.4	1.9	20.0	14.0
Weight	kg.	10	61.8	9.6	81.0	49.9
Height, Max. stretch	cm.	10	166.0	6.0	171.8	152.0
Trunk length	cm.	10	49.7	2.4	53.9	46.5
Biacromial breadth	cm.	10	37.7	1.0	39.1	35.8
Biiliocristal breadth	cm.	10	27.8	1.7	31.0	25.1
Upper plus lower arm length, right	cm.	10	55.2	3.5	61.2	48.6
Total leg length, right	cm.	10	77.3	5.2	84.3	68.7
Sum 3 skinfolds	mm.	10	36.3	8.4	50.8	22.5
Height-Weight ratio		10	12.90	.38	13.27	12.32
Endomorphy		10	3.7	.86	5.0	2.0
Mesomorphy		10	4.1	.70	5.5	3.0
Ectomorphy		10	2.6	1.02	3.5	1.0

TABLE 3.39. FREQUENCIES OF SOMATOTYPES IN SWIMMING, DIVING, CANOEING, AND GYMNASTICS (WOMEN) ACCORDING TO COMPONENT DOMINANCE

Somatotype	Swimming		Diving		Canoeing		Gymnastics	
Group	N	%	N	%	N	%	N	%
Endomorphy	1	4						
Endo-Meso	5	18			3	75	1	5
Mesomorphy	2	7	1	14	1	25	6	29
Ecto-Meso	2	7					1	5
Ectomorphy	1	4						
Central	17	61	6	86			13	62
Totals	28	101	7	100	4	100	21	101

TABLE 3.40. ANALYSIS OF DIFFERENCES BETWEEN GROUPS - SWIMMING (WOMEN)

Variable	F-ratio	Between group differences			
Height	0.93	Breast	Back	Free	Medley
Weight	1.70	Free	Breast	Back	Medley
Endomorphy	2.65	Breast	Free	Back	Medley
Mesomorphy	1.72	Free	Medley	Breast	Back
Ectomorphy	3.78*	Medley	Breast	Back	Free
Biacromiale	0.90	Back	Breast	Free	Medley
Biiliocristale	1.09	Free	Breast	Back	Medley
Arm length	0.17	Back	Medley	Free	Breast
Leg length	1.44	Breast	Back	Medley	Free
Trunk length	0.08	Free	Breast	Back	Medley

* $F_{.05}$ (3.23) = 3.03 ** $F_{.01}$ (3.23) = 4.76

TABLE 3.41. DESCRIPTIVE ANALYSIS FOR DIVING, MEN

Variable	Units	N	Mean	S.D.	Max.	Min.
Age	yr.	16	21.3	3.7	30.0	16.0
Weight	kg.	16	65.5	5.0	74.2	58.1
Height, Max. stretch	cm.	16	172.1	5.5	181.6	161.6
Trunk length	cm.	16	51.3	2.2	55.0	46.9
Biacromial breadth	cm.	16	40.2	2.2	43.3	34.0
Biiliocristal breadth	cm.	16	27.3	1.4	30.8	25.7
Upper plus lower arm length, right	cm.	16	58.5	2.5	62.6	53.9
Total leg length, right	cm.	16	79.0	3.6	85.6	74.0
Sum 3 skinfolds	mm.	16	20.0	3.9	30.6	15.6
Height-Weight ratio		16	12.92	.29	13.50	12.45
Endomorphy		16	1.9	.49	3.0	1.5
Mesomorphy		16	5.4	.74	7.0	4.5
Ectomorphy		16	2.7	.70	4.0	1.5

TABLE 3.42. DESCRIPTIVE ANALYSIS FOR DIVING, WOMEN

Variable	Units	N	Mean	S.D.	Max.	Min.
Age	yr.	7	21.1	7.6	38.0	16.0
Weight	kg.	7	52.3	3.9	59.7	47.7
Height, Max. stretch	cm.	7	160.4	2.9	165.3	157.2
Trunk length	cm.	7	47.9	2.0	51.3	45.3
Biacromial breadth	cm.	7	36.5	1.5	38.0	34.2
Biiliocristal breadth	cm.	7	26.5	1.7	29.3	24.8
Upper plus lower arm length, right	cm.	7	53.3	2.6	56.5	49.6
Total leg length, right	cm.	7	74.4	2.5	79.1	71.8
Sum 3 skinfolds	mm.	7	28.0	6.1	37.3	21.8
Height-Weight ratio		7	12.99	.24	13.33	12.75
Endomorphy		7	2.9	.80	4.0	2.0
Mesomorphy		7	4.0	.64	5.0	3.0
Ectomorphy		7	2.9	.48	3.5	2.5

TABLE 3.43. FREQUENCIES OF SOMATOTYPES ACCORDING TO COMPONENT DOMINANCE

Somatotype Group	Diving Men		Water Polo		Modern Pentathlon		Basketball		Rowing		Canoeing Men		Cycling	
	N	%	N	%	N	%	N	%	N	%	N	%	N	%
Endomorphy														
Endo-Meso			27	38	2	8	1	2	7	8	2	4	4	4
Mesomorphy	11	69	26	37	15	63	15	24	56	65	30	61	50	50
Ecto-Meso	5	31	9	13	6	25	25	40	16	19	14	29	38	38
Ectomorphy							2	3			1	2		
Central			9	13	1	4	20	32	7	8	2	4	8	8
Totals	16	100	71	101	24	100	63	101	86	100	49	100	100	100

TABLE 3.44. DESCRIPTIVE ANALYSIS FOR WATER POLO

Variable	Units	N	Mean	S.D.	Max.	Min.
Age	yr.	71	22.9	4.2	37.0	16.0
Weight	kg.	71	77.8	8.5	102.8	55.3
Height, Max. stretch	cm.	71	179.9	6.9	196.0	166.2
Trunk length	cm.	71	54.9	2.9	62.4	49.1
Biacromial breadth	cm.	71	42.2	1.8	47.5	39.1
Biiliocristal breadth	cm.	71	28.9	1.7	33.0	23.6
Upper plus lower arm length, right	cm.	71	60.4	2.7	66.4	54.7
Total leg length,right	cm.	70	82.6	4.1	94.0	75.2
Sum 3 skinfolds	mm.	71	29.3	11.1	70.3	14.2
Height-Weight ratio		71	12.75	.37	13.50	11.90
Endomorphy		71	2.9	1.18	6.5	1.0
Mesomorphy		71	5.3	.85	7.5	4.0
Ectomorphy		71	2.3	.83	4.0	.5

TABLE 3.45. DESCRIPTIVE ANALYSIS FOR MODERN PENTATHLON

Variable	Units	N	Mean	S.D.	Max.	Min.
Age	yr.	24	24.9	3.1	33.0	19.0
Weight	kg.	24	69.6	5.1	84.7	56.7
Height, Max. stretch	cm.	24	174.7	5.0	187.4	167.1
Trunk length	cm.	24	52.7	2.7	62.5	48.2
Biacromial breadth	cm.	24	40.8	1.9	45.3	37.2
Biiliocristal breadth	cm.	24	28.5	1.3	31.4	26.0
Upper plus lower arm length, right	cm.	23	59.2	2.6	62.5	52.7
Total leg length,right	cm.	23	80.6	3.8	88.4	74.7
Sum 3 skinfolds	mm.	24	21.2	6.1	41.2	13.6
Height-Weight ratio		24	12.83	.29	13.41	12.25
Endomorphy		24	2.0	.78	4.5	1.0
Mesomorphy		24	5.3	.75	6.5	4.0
Ectomorphy		24	2.4	.74	4.0	1.0

TABLE 3.46. DESCRIPTIVE ANALYSIS FOR BASKETBALL

Variable	Units	N	Mean	S.D.	Max.	Min.
Age	yr.	63	24.0	4.1	38.0	18.0
Weight	kg.	63	79.7	9.9	109.1	60.9
Height, Max. stretch	cm.	63	189.1	8.5	212.0	171.3
Trunk length	cm.	63	56.4	2.8	64.3	50.7
Biacromial breadth	cm.	63	42.5	2.3	48.5	35.5
Biiliocristal breadth	cm.	63	29.5	2.4	34.0	23.9
Upper plus lower arm length, right	cm.	63	64.3	3.4	74.1	56.4
Total leg length,right	cm.	63	89.3	5.8	106.0	77.6
Sum 3 skinfolds	mm.	63	21.2	6.0	53.5	13.0
Height-Weight ratio		63	13.26	.36	14.39	12.42
Endomorphy		63	2.0	.72	5.5	1.0
Mesomorphy		63	4.3	.86	6.5	2.5
Ectomorphy		63	3.5	.89	6.0	1.5

TABLE 3.47. DESCRIPTIVE ANALYSIS FOR ROWING, TOTAL GROUP

Variable	Units	N	Mean	S.D.	Max.	Min.
Age	yr.	86	24.3	3.3	40.0	18.0
Weight	kg.	85	82.6	7.4	106.5	65.6
Height, Max. stretch	cm.	85	185.1	5.9	198.4	172.1
Trunk length	cm.	86	55.4	2.7	62.5	47.1
Biacromial breadth	cm.	86	42.5	1.8	48.1	38.6
Biiliocristal breadth	cm.	86	30.0	1.9	34.0	26.8
Upper plus lower arm length, right	cm.	86	62.6	2.8	69.9	57.2
Total leg length,right	cm.	86	86.2	4.0	96.7	77.3
Sum 3 skinfolds	mm.	85	21.8	5.0	36.1	13.4
Height-Weight ratio		85	12.83	.32	13.70	12.09
Endomorphy		85	2.1	.64	4.0	1.0
Mesomorphy		85	5.3	.91	7.0	3.0
Ectomorphy		85	2.4	.78	4.5	1.0

TABLE 3.48. DESCRIPTIVE ANALYSIS FOR ROWING, SINGLE AND DOUBLE SCULLS

Variable	Units	N	Mean	S.D.	Max.	Min.
Age	yr.	16	24.6	2.8	31.0	20.0
Weight	kg.	16	80.6	4.9	91.7	74.4
Height, Max. stretch	cm.	16	184.3	6.4	195.6	174.6
Trunk length	cm.	16	55.1	2.3	59.9	51.6
Biacromial breadth	cm.	16	42.2	1.7	44.7	39.0
Biiliocristal breadth	cm.	16	29.6	1.4	31.8	27.5
Upper plus lower arm length, right	cm.	16	62.2	2.8	66.0	57.2
Total leg length,right	cm.	16	85.9	4.2	93.5	78.4
Sum 3 skinfolds	mm.	16	20.6	4.3	28.2	14.7
Height-Weight ratio		16	12.87	.36	13.60	12.21
Endomorphy		16	1.9	.60	3.0	1.0
Mesomorphy		16	5.3	1.22	7.0	3.5
Ectomorphy		16	2.5	.90	4.5	1.0

TABLE 3.49. DESCRIPTIVE ANALYSIS FOR ROWING, PAIRS

Variable	Units	N	Mean	S.D.	Max.	Min.
Age	yr.	20	25.1	4.5	40.0	19.0
Weight	kg.	20	81.8	5.3	90.0	71.9
Height, Max. stretch	cm.	20	184.0	5.0	190.5	172.1
Trunk length	cm.	20	55.3	2.8	61.0	49.0
Biacromial breadth	cm.	20	42.4	1.8	44.7	39.0
Biiliocristal breadth	cm.	20	30.0	1.4	33.2	27.7
Upper plus lower arm length, right	cm.	20	62.2	2.6	66.4	57.5
Total leg length,right	cm.	20	85.7	3.7	90.7	77.8
Sum 3 skinfolds	mm.	20	22.7	5.3	34.0	13.9
Height-Weight ratio		20	12.79	.30	13.45	12.09
Endomorphy		20	2.1	.69	3.5	1.0
Mesomorphy		20	5.4	.77	7.0	4.0
Ectomorphy		20	2.3	.73	4.0	1.0

TABLE 3.50. DESCRIPTIVE ANALYSIS FOR ROWING, FOURS

Variable	Units	N	Mean	S.D.	Max.	Min.
Age	yr.	23	24.3	3.0	29.0	18.0
Weight	kg.	23	81.7	7.2	106.5	70.9
Height, Max. stretch	cm.	23	184.5	5.6	198.4	174.0
Trunk length	cm.	23	54.6	1.9	59.1	51.4
Biacromial breadth	cm.	23	42.7	2.0	48.1	39.7
Biiliocristal breadth	cm.	23	30.1	1.6	33.4	26.9
Upper plus lower arm length, right	cm.	23	62.8	2.7	69.9	58.6
Total leg length,right	cm.	23	86.2	4.0	94.4	77.3
Sum 3 skinfolds	mm.	23	21.2	5.8	36.1	13.4
Height-Weight ratio		23	12.83	.24	13.20	12.35
Endomorphy		23	2.1	.74	4.0	1.0
Mesomorphy		23	5.3	.78	7.0	4.0
Ectomorphy		23	2.4	.57	3.5	1.5

TABLE 3.51. DESCRIPTIVE ANALYSIS FOR ROWING, EIGHTS

Variable	Units	N	Mean	S.D.	Max.	Min.
Age	yr.	27	23.2	2.7	28.0	19.0
Weight	kg.	26	85.1	9.7	103.3	65.6
Height, Max. stretch	cm.	26	187.0	6.4	198.4	175.6
Trunk length	cm.	27	56.6	3.8	62.5	47.1
Biacromial breadth	cm.	27	42.4	1.8	46.6	38.6
Biiliocristal breadth	cm.	26	30.5	2.0	34.0	26.8
Upper plus lower arm length, right	cm.	27	63.1	3.0	69.2	59.1
Total leg length,right	cm.	27	86.8	4.2	96.7	80.5
Sum 3 skinfolds	mm.	26	22.2	4.4	30.4	15.6
Height-Weight ratio		26	12.84	.36	13.70	12.25
Endomorphy		26	2.2	.53	3.0	1.5
Mesomorphy		26	5.3	.95	6.5	3.0
Ectomorphy		26	2.5	.92	4.5	1.0

TABLE 3.52. ANALYSIS OF DIFFERENCES BETWEEN GROUPS - ROWING

Variable	F-ratio*	Between group differences			
Height	1.49	Fours	Pairs	Single Doubles	Eights
Weight	1.80	Fours	Single Doubles	Pairs	Eights
Endomorphy	0.76	Single Doubles	Fours	Eights	Pairs
Mesomorphy	0.10	Fours	Eights	Single Doubles	Pairs
Ectomorphy	0.22	Pairs	Eights	Fours	Single Doubles
Biacromiale	0.30	Fours	Single Doubles	Pairs	Eights
Biiliocristale	0.39	Single Doubles	Pairs	Fours	Eights
Arm length	0.55	Single Doubles	Pairs	Fours	Eights
Leg length	0.34	Pairs	Single Doubles	Fours	Eights
Trunk length	1.33	Single Doubles	Fours	Pairs	Eights

$^*F_{.05}(3,82) = 2.72$

TABLE 3.53. DESCRIPTIVE ANALYSIS FOR CANOEING, TOTAL GROUP, MEN

Variable	Units	N	Mean	S.D.	Max.	Min.
Age	yr.	49	24.2	4.3	38.0	18.0
Weight	kg.	49	74.4	7.7	90.4	60.2
Height, Max. stretch	cm.	49	178.5	7.9	197.3	160.0
Trunk length	cm.	49	53.4	3.3	63.8	46.8
Biacromial breadth	cm.	49	41.8	1.7	46.4	37.3
Biiliocristal breadth	cm.	48	29.1	1.8	33.1	26.2
Upper plus lower arm length, right	cm.	49	60.7	3.4	69.6	53.0
Total leg length, right	cm.	49	82.8	4.9	96.2	72.6
Sum 3 skinfolds	mm.	49	20.0	5.4	46.9	13.2
Height-Weight ratio		49	12.83	.40	13.85	11.80
Endomorphy		49	1.9	.67	5.0	1.0
Mesomorphy		49	5.5	.95	7.5	3.0
Ectomorphy		49	2.5	.94	5.0	.5

TABLE 3.54. DESCRIPTIVE ANALYSIS FOR CANOEING, MEN, KAYAK (1 MAN, 2 AND 4 MEN)

Variable	Units	N	Mean	S.D.	Max.	Min.
Age	yr.	34	25.0	4.3	38.0	19.0
Weight	kg.	34	75.9	8.0	90.4	63.3
Height, Max. stretch	cm.	34	180.3	7.5	197.3	167.3
Trunk length	cm.	34	54.2	3.4	63.8	49.8
Biacromial breadth	cm.	34	42.0	1.9	46.4	37.3
Biiliocristal breadth	cm.	34	29.3	1.8	33.1	26.4
Upper plus lower arm length, right	cm.	34	61.5	3.4	69.6	55.2
Total leg length,right	cm.	34	83.8	4.9	96.2	72.6
Sum 3 skinfolds	mm.	34	20.0	6.0	46.9	13.2
Height-Weight ratio		34	12.88	.42	13.85	11.80
Endomorphy		34	1.9	.74	5.0	1.0
Mesomorphy		34	5.4	.98	7.5	3.0
Ectomorphy		34	2.6	1.01	5.0	.5

TABLE 3.55. DESCRIPTIVE ANALYSIS FOR CANOEING, MEN, CANADIAN

Variable	Units	N	Mean	S.D.	Max.	Min
Age	yr.	15	22.3	3.7	31.0	18.0
Weight	kg.	15	71.0	5.8	79.6	60.2
Height, Max. stretch	cm.	15	174.2	7.4	185.9	160.0
Trunk length	cm.	15	51.6	2.1	55.5	46.8
Biacromial breadth	cm.	15	41.4	.9	43.1	39.8
Biiliocristal breadth	cm.	14	28.5	1.2	30.6	26.2
Upper plus lower arm length, right	cm.	15	58.8	2.6	62.4	53.0
Total leg length,right	cm.	15	80.4	4.2	89.8	74.7
Sum 3 skinfolds	mm.	15	20.1	3.8	28.6	16.1
Height-Weight ratio		15	12.71	.34	13.50	12.34
Endomorphy		15	1.9	.49	3.0	1.5
Mesomorphy		15	5.6	.88	6.5	3.5
Ectomorphy		15	2.3	.75	4.0	1.5

TABLE 3.56. ANALYSIS OF DIFFERENCES BETWEEN KAYAK (N=34) AND CANADIAN (N=15) STYLE CANOEISTS (MEN)

Variable	Kayak Mean	Canadian Mean	Mean Difference	F-ratio
Height (cm)	180.4	171.2	9.2	7.04*
Weight (kg)	75.9	71.1	4.8	4.51*
Skinfolds (mm) (Endomorphy)	20.0	20.1	-0.1	0.004
Mesomorphy	5.4	5.6	-0.2	0.72
H. W. R. (Ectomorphy)	12.88	12.72	0.16	1.57
Acromiale (cm)	42.0	41.4	0.6	1.19
Biiliocristale (cm)	29.3	28.5	0.8	2.21
Arm length (cm)	61.5	58.8	2.7	7.43**
Leg length (cm)	83.8	80.4	3.4	5.33*
Trunk length (cm)	54.2	51.6	2.6	7.56**

* $F_{.05}(1,47) = 4.05$ ** $F_{.01}(1,47) = 7.20$

TABLE 3.57. DESCRIPTIVE ANALYSIS FOR CANOEING (WOMEN)

Variable	Units	N	Mean	S.D.	Max.	Min.
Age	yr.	4	22.0	2.7	25.0	18.0
Weight	kg.	4	61.0	2.8	63.6	58.0
Height, Max. stretch	cm.	4	163.1	3.9	168.0	159.4
Trunk length	cm.	4	49.3	1.4	51.3	48.3
Biacromial breadth	cm.	4	38.0	0.4	38.4	37.6
Biiliocristal breadth	cm.	4	27.8	1.1	29.0	26.5
Upper plus lower arm length, right	cm.	4	54.2	2.8	58.2	51.8
Total leg length,right	cm.	4	75.6	3.7	80.7	71.8
Sum 3 skinfolds	mm.	4	35.4	9.1	44.4	22.9
Height-Weight ratio		4	12.53	0.16	12.77	12.45
Endomorphy		4	3.5	0.94	4.5	2.0
Mesomorphy		4	5.3	0.56	6.0	4.5
Ectomorphy		4	1.8	0.43	2.5	1.5

TABLE 3.58. DESCRIPTIVE ANALYSIS FOR CYCLING, TOTAL GROUP

Variable	Units	N	Mean	S.D.	Max.	Min.
Age	yr.	104	23.6	3.4	32.0	17.0
Weight	kg.	100	68.9	7.7	88.6	52.0
Height, Max. stretch	cm.	100	174.9	6.5	189.4	161.0
Trunk length	cm.	102	53.5	2.6	60.0	46.9
Biacromial breadth	cm.	101	39.9	2.0	44.8	33.1
Biiliocristal breadth	cm.	102	28.1	1.7	32.4	24.3
Upper plus lower arm length, right	cm.	101	58.7	2.9	65.8	51.5
Total leg length,right	cm.	102	80.0	4.3	89.6	69.6
Sum 3 skinfolds	mm.	101	19.1	3.8	34.3	13.3
Height-Weight ratio		100	12.91	.33	13.63	12.10
Endomorphy		100	1.8	.49	3.5	1.0
Mesomorphy		100	5.0	.83	7.0	3.0
Ectomorphy		100	2.7	.79	4.5	1.0

TABLE 3.59. DESCRIPTIVE ANALYSIS FOR CYCLING, SPRINTING (1,000 m TIME TRIAL, SCRATCH SPRINT)

Variable	Units	N	Mean	S.D.	Max.	Min.
Age	yr.	15	22.7	3.0	29.0	18.0
Weight	kg.	14	69.8	9.1	88.6	56.7
Height, Max. stretch	cm.	14	174.0	5.7	187.4	164.5
Trunk length	cm.	15	53.3	2.2	57.7	49.5
Biacromial breadth	cm.	15	40.1	1.7	42.6	37.5
Biiliocristal breadth	cm.	15	27.9	2.0	32.4	24.3
Upper plus lower arm length, right	cm.	15	58.2	2.9	63.2	53.3
Total leg length,right	cm.	14	79.4	3.8	85.7	73.9
Sum 3 skinfolds	mm.	14	19.5	3.2	28.5	15.8
Height-Weight ratio		14	12.81	.30	13.20	12.30
Endomorphy		14	1.8	.42	3.0	1.5
Mesomorphy		14	5.2	.80	7.0	4.0
Ectomorphy		14	2.4	.68	3.5	1.0

TABLE 3.60. DESCRIPTIVE ANALYSIS FOR CYCLING, PURSUIT (4,000 m INDIVIDUAL, 4000 m TIME TRIAL)

Variable	Units	N	Mean	S.D.	Max.	Min.
Age	yr.	21	22.9	3.5	32.0	18.0
Weight	kg.	19	68.8	9.5	86.7	55.4
Height, Max. stretch	cm.	19	174.6	6.7	186.3	163.4
Trunk length	cm.	19	53.9	2.3	59.9	49.5
Biacromial breadth	cm.	18	39.9	2.2	44.8	36.8
Biiliocristal breadth	cm.	19	27.6	1.9	30.6	24.8
Upper plus lower arm length, right	cm.	19	57.7	3.0	62.0	51.5
Total leg length,right	cm.	19	79.9	4.6	87.2	69.6
Sum 3 skinfolds	mm.	19	18.9	4.2	32.6	13.4
Height-Weight ratio		19	12.90	.42	13.63	12.10
Endomorphy		19	1.8	.53	3.5	1.0
Mesomorphy		19	5.1	1.05	7.0	3.0
Ectomorphy		19	2.6	1.03	4.5	1.0

TABLE 3.61. DESCRIPTIVE ANALYSIS FOR CYCLING, ROAD (INDIVIDUAL AND TEAM ROAD RACE)

Variable	Units	N	Mean	S.D.	Max.	Min.
Age	yr.	68	24.0	3.4	32.0	17.0
Weight	kg.	67	68.7	6.9	83.6	52.0
Height, Max. stretch	cm.	67	175.1	6.7	189.4	161.0
Trunk length	cm.	68	53.5	2.7	60.0	46.9
Biacromial breadth	cm.	68	39.8	2.0	43.4	33.1
Biiliocristal breadth	cm.	68	28.3	1.6	32.1	25.3
Upper plus lower arm length, right	cm.	67	59.1	2.8	65.0	53.1
Total leg length,right	cm.	69	80.3	4.1	89.6	70.8
Sum 3 skinfolds	mm.	68	19.1	3.8	34.3	13.3
Height-Weight ratio		67	12.94	.30	13.55	12.25
Endomorphy		67	1.8	.50	3.5	1.0
Mesomorphy		67	4.9	.77	7.0	3.5
Ectomorphy		67	2.7	.74	4.0	1.0

TABLE 3.62. ANALYSIS OF DIFFERENCES BETWEEN GROUPS CYCLING

Variable	F-ratio *	Between group differences		
Height	0.20	Sprint	Pursuit	Road
Weight	0.11	Road	Pursuit	Sprint
Endomorphy	0.10	Pursuit	Road	Sprint
Mesomorphy	0.87	Road	Pursuit	Sprint
Ectomorphy	0.96	Sprint	Pursuit	Road
Biacromiale	0.46	Pursuit	Road	Sprint
Biiliocristale	1.23	Pursuit	Sprint	Road
Arm length	2.24	Pursuit	Sprint	Road
Leg length	0.75	Pursuit	Sprint	Road
Trunk length	0.21	Sprint	Road	Pursuit

* $F_{.05}$ (2,97) = 3.09

TABLE 3.63. DESCRIPTIVE ANALYSIS FOR GYMNASTICS, MEN

Variable	Units	N	Mean	S.D.	Max.	Min.
Age	yr.	28	23.6	3.8	31.0	18.0
Weight	kg.	28	61.5	5.5	69.7	50.8
Height, Max. stretch	cm.	28	167.4	5.7	178.6	157.8
Trunk length	cm.	27	50.7	2.5	55.5	46.3
Biacromial breadth	cm.	27	40.0	1.6	43.5	37.3
Biiliocristal breadth	cm.	27	26.7	1.3	28.8	23.5
Upper plus lower arm length, right	cm.	26	55.8	2.6	60.6	51.8
Total leg length,right	cm.	27	76.7	5.0	91.1	67.6
Sum 3 skinfolds	mm.	28	16.3	2.0	20.3	12.6
Height-Weight ratio		28	12.82	.35	13.41	11.95
Endomorphy		28	1.4	.27	2.0	1.0
Mesomorphy		28	5.9	.82	8.0	4.5
Ectomorphy		28	2.4	.86	4.0	.5

TABLE 3.64. DESCRIPTIVE ANALYSIS FOR GYMNASTICS, WOMEN

Variable	Units	N	Mean	S.D.	Max.	Min.
Age	yr.	21	17.8	3.7	26.0	13.0
Weight	kg.	21	49.8	4.5	60.4	43.5
Height, Max. stretch	cm.	21	156.9	5.1	165.4	150.0
Trunk length	cm.	21	47.5	2.2	51.6	43.3
Biacromial breadth	cm.	20	35.7	1.7	39.1	34.8
Biiliocristal breadth	cm.	21	25.6	1.3	27.7	22.6
Upper plus lower arm length, right	cm.	21	52.2	2.0	56.2	47.7
Total leg length,right	cm.	20	72.2	2.8	78.3	67.7
Sum 3 skinfolds	mm.	21	26.8	6.2	42.3	17.5
Height-Weight ratio		21	12.98	.22	13.45	12.56
Endomorphy		21	2.7	.70	4.5	1.5
Mesomorphy		21	4.2	.46	5.0	3.0
Ectomorphy		21	2.8	.53	4.0	2.0

TABLE 3.65. FREQUENCIES OF SOMATOTYPES ACCORDING TO COMPONENT DOMINANCE

Somatotype Group	Gymnastics (Men)		Wrestling Freestyle		Wrestling Greco-Roman		Wrestling Total		Boxing		Weight lifting	
	N	%	N	%	N	%	N	%	N	%	N	%
Endomorphy												
Endo-Meso			16	33	12	30	28	31	8	6	22	38
Mesomorphy	14	50	30	61	27	68	57	64	83	59	35	60
Ecto-Meso	14	50	3	6	1	3	4	5	45	32	1	2
Ectomorphy									1	1		
Central									4	3		
Totals	28	100	49	100	40	101	89	100	141	101	58	100

TABLE 3.66. DESCRIPTIVE ANALYSIS FOR WRESTLING FREESTYLE, TOTAL GROUP

Variable	Units	N	Mean	S.D.	Max.	Min.
Age	yr.	50	25.8	5.1	37.0	18.0
Weight	kg.	49	70.6	12.4	106.3	51.7
Height, Max. stretch	cm.	49	169.3	8.7	195.4	150.4
Trunk length	cm.	49	52.2	3.1	58.7	44.1
Biacromial breadth	cm.	49	40.4	2.6	45.5	34.1
Biiliocristal breadth	cm.	49	27.9	1.8	34.7	24.4
Upper plus lower arm length, right	cm.	49	56.6	4.3	66.5	48.4
Total leg length, right	cm.	48	77.6	6.4	97.3	63.6
Sum 3 skinfolds	mm.	49	23.0	7.6	49.7	12.9
Height-Weight ratio		49	12.44	.38	13.64	11.53
Endomorphy		49	2.2	.85	5.0	1.0
Mesomorphy		49	6.3	1.07	9.0	4.0
Ectomorphy		49	1.6	.80	4.5	.5

TABLE 3.67. DESCRIPTIVE ANALYSIS FOR WRESTLING GRECO-ROMAN, TOTAL GROUP

Variable	Units	N	Mean	S.D.	Max.	Min.
Age	yr.	40	25.9	4.7	36.0	17.0
Weight	kg.	40	72.2	13.7	124.5	53.4
Height, Max. stretch	cm.	40	170.4	8.4	189.8	156.6
Trunk length	cm.	40	52.2	3.0	58.6	45.8
Biacromial breadth	cm.	40	40.5	2.5	45.1	36.2
Biiliocristal breadth	cm.	40	28.2	2.0	34.7	24.9
Upper plus lower arm length, right	cm.	39	57.2	3.3	63.8	52.4
Total leg length, right	cm.	40	77.7	5.7	91.8	69.6
Sum 3 skinfolds	mm.	40	24.4	11.0	59.9	14.3
Height-Weight ratio		40	12.43	.37	13.60	11.42
Endomorphy		40	2.4	1.20	6.0	1.0
Mesomorphy		40	6.4	.91	8.5	4.5
Ectomorphy		40	1.6	.77	4.5	.5

TABLE 3.68. ANALYSIS OF DIFFERENCES BETWEEN FREE-STYLE (N=48) AND GRECO-ROMAN (N=39) WRESTLERS

Variable	F-S Mean	G-R Mean	Mean Difference	F-ratio*
Height (cm)	169.1	170.3	-1.2	0.44
Weight (kg)	70.4	72.3	-1.9	0.44
Skinfolds (mm) (Endomorphy)	23.1	24.6	-1.5	0.54
Mesomorphy	6.4	6.5	-0.1	0.26
H. W. R. (Ectomorphy)	12.44	12.42	0.02	0.04
Acromiale (cm)	40.3	40.6	-0.3	0.18
Biiliocristale (cm)	27.9	28.2	-0.3	0.57
Arm length (cm)	56.5	57.2	-0.7	0.72
Leg length (cm)	77.6	77.6	0.0	0.003
Trunk length (cm)	52.2	52.3	-0.1	0.03

* $F_{.05}$ (1,85) = 3.96

TABLE 3.69. DESCRIPTIVE ANALYSIS FOR WRESTLING, FREE STYLE AND GRECO-ROMAN

Variable	Units	N	Mean	S.D.	Max.	Min.
Age	yr.	90	25.8	4.9	37.0	17.0
Weight	kg.	89	71.3	13.0	124.5	51.7
Height, Max. stretch	cm.	89	170.0	8.5	195.4	150.4
Trunk length	cm.	89	52.2	3.1	58.7	44.1
Biacromial breadth	cm.	89	40.4	2.4	45.5	34.1
Biiliocristal breadth	cm.	89	28.1	1.9	34.7	24.4
Upper plus lower arm length, right	cm.	88	56.9	3.9	66.5	48.4
Total leg length,right	cm.	88	77.7	6.1	97.3	63.6
Sum 3 skinfolds	mm.	89	23.7	9.3	59.9	12.9
Height-Weight ratio		89	12.44	.37	13.64	11.42
Endomorphy		89	2.3	1.01	6.0	1.0
Mesomorphy		89	6.4	1.00	9.0	4.0
Ectomorphy		89	1.6	.79	4.5	.5

TABLE 3.70. DESCRIPTIVE ANALYSIS FOR WRESTLING, FLY AND BANTAM WEIGHTS

Variable	Units	N	Mean	S.D.	Max.	Min.
Age	yr.	23	24.3	4.0	36.0	18.0
Weight	kg.	22	57.8	4.3	70.0	51.7
Height, Max. stretch	cm.	22	161.4	5.7	175.5	150.4
Trunk length	cm.	22	50.0	2.3	53.7	44.1
Biacromial breadth	cm.	22	38.0	1.6	40.5	34.1
Biiliocristal breadth	cm.	22	26.5	1.3	28.4	24.4
Upper plus lower arm length, right	cm.	22	54.1	3.3	62.3	48.4
Total leg length,right	cm.	22	72.5	4.7	84.7	63.6
Sum 3 skinfolds	mm.	22	19.0	4.0	28.4	12.9
Height-Weight ratio		22	12.65	1.39	13.64	12.14
Endomorphy		22	1.8	.50	3.0	1.0
Mesomorphy		22	5.9	.98	7.5	4.0
Ectomorphy		22	2.1	.95	4.5	1.0

TABLE 3.71. DESCRIPTIVE ANALYSIS FOR WRESTLING, FEATHER AND LIGHT WEIGHTS

Variable	Units	N	Mean	S.D.	Max.	Min.
Age	yr.	42	26.8	5.6	36.0	17.0
Weight	kg.	42	68.7	3.9	76.1	59.4
Height, Max. stretch	cm.	42	168.3	3.7	176.6	160.2
Trunk length	cm.	42	52.1	2.5	58.0	46.9
Biacromial breadth	cm.	42	40.4	1.7	42.9	35.4
Biiliocristal breadth	cm.	42	27.7	1.3	30.2	24.6
Upper plus lower arm length, right	cm.	41	56.0	3.0	62.4	48.5
Total leg length,right	cm.	42	77.5	4.8	91.1	68.2
Sum 3 skinfolds	mm.	42	22.5	7.8	49.7	14.6
Height-Weight ratio		42	12.44	.29	13.18	11.83
Endomorphy		42	2.2	.88	5.0	1.0
Mesomorphy		42	6.3	.83	8.5	4.5
Ectomorphy		42	1.6	.66	3.5	.5

TABLE 3.72. DESCRIPTIVE ANALYSIS FOR WRESTLING, WELTER AND MIDDLE WEIGHTS

Variable	Units	N	Mean	S.D.	Max.	Min.
Age	yr.	18	25.7	4.0	37.0	20.0
Weight	kg.	18	82.7	4.5	90.9	74.4
Height, Max. stretch	cm.	18	177.3	4.9	183.7	167.5
Trunk length	cm.	18	54.0	2.4	58.7	50.1
Biacromial breadth	cm.	18	42.2	1.5	45.5	40.2
Biiliocristal breadth	cm.	18	29.5	1.0	31.2	27.7
Upper plus lower arm length, right	cm.	18	60.0	2.1	63.1	55.7
Total leg length,right	cm.	17	81.4	4.1	91.8	76.5
Sum 3 skinfolds	mm.	18	27.5	8.6	48.3	17.3
Height-Weight ratio		18	12.31	.39	12.95	11.53
Endomorphy		18	2.7	.93	5.0	1.5
Mesomorphy		18	6.9	1.19	9.0	5.0
Ectomorphy		18	1.3	.66	2.5	.5

TABLE 3.73. DESCRIPTIVE ANALYSIS FOR WRESTLING, LIGHT HEAVY AND HEAVY WEIGHTS

Variable	Units	N	Mean	S.D.	Max.	Min.
Age	yr.	7	24.6	4.8	31.0	20.0
Weight	kg.	7	100.4	12.2	124.5	88.0
Height, Max. stretch	cm.	7	186.0	6.0	195.4	177.0
Trunk length	cm.	7	56.2	2.5	58.6	51.5
Biacromial breadth	cm.	7	44.2	1.0	45.1	42.4
Biiliocristal breadth	cm.	7	31.2	2.1	34.7	29.0
Upper plus lower arm length, right	cm.	7	62.7	2.6	66.5	58.4
Total leg length,right	cm.	7	86.1	6.3	97.3	77.0
Sum 3 skinfolds	mm.	7	35.2	16.9	59.9	21.9
Height-Weight ratio		7	12.08	.35	12.45	11.42
Endomorphy		7	3.6	1.69	6.0	2.0
Mesomorphy		7	7.1	.61	8.0	6.5
Ectomorphy		7	1.0	.41	1.5	.5

TABLE 3.74. ANALYSIS OF DIFFERENCES BETWEEN GROUPS - WRESTLING

Variable	F-ratio	Between group differences			
Height	68.62**	Fly-Bantam	Feather-Light	Welter-Middle	Light Heavy-Heavy
Endomorphy	8.41**	Fly-Bantam	Feather-Light	Welter-Middle	Light Heavy-Heavy
Mesomorphy	4.82**	Fly-Bantam	Feather-Light	Welter-Middle	Light Heavy-Heavy
Ectomorphy	6.09**	Light Heavy Heavy	Welter-Middle	Feather-Light	Fly-Bantam
Biacromiale	36.18**	Fly-Bantam	Feather-Light	Welter-Middle	Light Heavy-Heavy
Biiliocristale	30.67**	Fly-Bantam	Feather-Light	Welter-Middle	Light Heavy-Heavy
Arm length	24.41**	Fly-Bantam	Feather-Light	Welter-Middle	Light Heavy-Heavy
Leg length	19.53**	Fly-Bantam	Feather-Light	Welter-Middle	Light Heavy-Heavy
Trunk length	17.66**	Fly-Bantam	Feather-Light	Welter-Middle	Light Heavy-Heavy

* $F_{.05}$ (3.84) = 2.72 ** $F_{.01}$ (3.84) = 4.03

TABLE 3.75. DESCRIPTIVE ANALYSIS FOR BOXING, TOTAL GROUP

Variable	Units	N	Mean	S.D.	Max.	Min.
Age	yr.	142	22.9	3.4	35.0	17.0
Weight	kg.	141	63.1	9.5	93.7	47.5
Height, Max. stretch	cm.	141	169.3	7.4	189.0	153.8
Trunk length	cm	141	51.3	2.8	59.9	45.1
Biacromial breadth	cm.	141	39.7	2.1	47.5	35.6
Biiliocristal breadth	cm.	140	27.1	1.8	31.4	22.7
Upper plus lower arm length, right	cm.	141	57.3	3.5	66.3	50.3
Total leg length, right	cm.	139	78.1	5.5	96.5	69.2
Sum 3 skinfolds	mm.	142	20.0	5.2	51.0	12.8
Height-Weight ratio		142	12.91	.38	14.20	11.82
Endomorphy		142	1.9	.62	5.0	1.0
Mesomorphy		142	5.3	.88	8.0	3.0
Ectomorphy		142	2.6	.88	6.0	.5

TABLE 3.76. DESCRIPTIVE ANALYSIS FOR BOXING, LIGHT FLY AND FLY WEIGHTS

Variable	Units	N	Mean	S.D.	Max.	Min.
Age	yr.	29	21.6	3.1	28.0	17.0
Weight	kg.	29	52.5	3.3	65.9	47.5
Height, Max. stretch	cm.	29	161.7	4.4	175.9	153.8
Trunk length	cm.	29	49.1	1.9	53.2	45.7
Biacromial breadth	cm.	29	37.7	1.2	40.3	35.6
Biiliocristal breadth	cm.	29	25.5	1.3	29.5	22.7
Upper plus lower arm length, right	cm.	28	55.5	3.6	66.3	50.5
Total leg length, right	cm.	29	74.1	3.0	82.0	69.2
Sum 3 skinfolds	mm.	29	17.7	3.0	25.5	12.8
Height-Weight ratio		29	13.09	.26	13.52	12.55
Endomorphy		29	1.6	.36	2.5	1.0
Mesomorphy		29	4.9	.64	6.0	3.5
Ectomorphy		29	3.0	.63	4.0	2.0

TABLE 3.77. DESCRIPTIVE ANALYSIS FOR BOXING, BANTAM AND FEATHER WEIGHTS

Variable	Units	N	Mean	S.D.	Max.	Min.
Age	yr.	32	22.5	2.9	30.0	18.0
Weight	kg.	32	57.0	2.4	61.3	52.7
Height, Max. stretch	cm.	32	166.1	5.3	177.9	154.8
Trunk length	cm.	32	50.2	2.3	54.0	45.1
Biacromial breadth	cm.	32	38.9	1.3	41.3	35.9
Biiliocristal breadth	cm.	31	26.5	1.4	29.4	23.8
Upper plus lower arm length, right	cm.	32	55.8	2.8	63.6	50.6
Total leg length, right	cm.	31	76.0	4.1	87.0	69.2
Sum 3 skinfolds	mm.	32	18.0	2.4	24.4	13.2
Height-Weight ratio		32	13.07	.39	14.20	12.20
Endomorphy		32	1.7	.35	2.5	1.0
Mesomorphy		32	4.9	.90	6.5	3.0
Ectomorphy		32	3.0	.94	6.0	1.0

TABLE 3.78. DESCRIPTIVE ANALYSIS FOR BOXING, LIGHT AND LIGHT WELTER WEIGHTS

Variable	Units	N	Mean	S.D.	Max.	Min.
Age	yr.	33	23.1	3.4	31.0	17.0
Weight	kg.	33	63.1	2.6	68.2	59.4
Height, Max. stretch	cm.	33	169.5	4.1	178.4	163.4
Trunk length	cm.	33	51.9	2.4	56.2	46.5
Biacromial breadth	cm.	33	39.8	1.9	47.5	37.7
Biiliocristal breadth	cm.	33	27.3	1.2	30.6	24.5
Upper plus lower arm length, right	cm.	33	56.9	2.6	61.4	52.2
Total leg length,right	cm.	33	77.7	4.2	87.3	69.5
Sum 3 skinfolds	mm.	33	19.8	3.6	30.6	14.4
Height-Weight ratio		33	12.90	.33	13.65	12.34
Endomorphy		33	1.8	.48	3.0	1.0
Mesomorphy		33	5.4	.75	7.5	4.0
Ectomorphy		33	2.6	.79	4.5	1.5

TABLE 3.79. DESCRIPTIVE ANALYSIS FOR BOXING, WELTER AND LIGHT MIDDLE WEIGHTS

Variable	Units	N	Mean	S.D.	Max.	Min.
Age	yr.	29	23.2	3.2	31.0	17.0
Weight	kg.	28	69.7	2.9	74.4	63.2
Height, Max. stretch	cm.	28	173.5	5.1	187.2	164.6
Trunk length	cm.	28	52.6	2.1	57.1	47.8
Biacromial breadth	cm.	28	41.0	1.7	44.3	37.0
Biiliocristal breadth	cm.	28	27.5	1.7	30.4	24.5
Upper plus lower arm length, right	cm.	29	58.6	2.8	64.2	51.5
Total leg length,right	cm.	28	80.6	5.1	96.5	73.0
Sum 3 skinfolds	mm.	29	21.4	5.0	35.7	13.8
Height-Weight ratio		29	12.76	.33	13.44	12.17
Endomorphy		29	2.1	.61	3.5	1.5
Mesomorphy		29	5.4	.79	7.5	4.0
Ectomorphy		29	2.3	.78	4.0	1.0

TABLE 3.80. DESCRIPTIVE ANALYSIS FOR BOXING, MIDDLE, LIGHT HEAVY, AND HEAVY WEIGHTS

Variable	Units	N	Mean	S.D.	Max.	Min.
Age	yr.	19	24.9	4.1	35.0	19.0
Weight	kg.	19	80.0	5.6	93.7	73.1
Height, Max. stretch	cm.	19	179.5	5.5	189.0	169.3
Trunk length	cm.	19	53.8	3.2	59.9	46.6
Biacromial breadth	cm.	19	42.1	1.7	45.5	38.2
Biiliocristal breadth	cm.	19	29.3	1.8	31.4	25.9
Upper plus lower arm length, right	cm.	19	61.0	3.2	65.8	54.4
Total leg length, right	cm.	19	84.9	5.2	92.9	73.7
Sum 3 skinfolds	mm.	19	24.8	9.1	51.0	15.1
Height-Weight ratio		19	12.59	.37	13.03	11.82
Endomorphy		19	2.4	1.00	5.0	1.5
Mesomorphy		19	6.0	1.01	8.0	4.5
Ectomorphy		19	1.9	.79	3.0	.5

TABLE 3.81. ANALYSIS OF DIFFERENCES BETWEEN GROUPS - BOXING

Variable	F-ratio	Between group differences					
Height	53.08**	Light Fly-Fly	Bantam-Feather	Light-Light Welter	Welter-Light Middle	Middle-Light	Heavy-Heavy
Endomorphy	9.26**	Light Fly-Fly	Bantam-Feather	Light-Light Welter	Welter-Light Middle	Middle-Light	Heavy-Heavy
Mesomorphy	7.29**	Light Fly-Fly	Bantam-Feather	Welter-Light Middle	Light-Light Welter	Middle-Light	Heavy-Heavy
Ectomorphy	9.22**	Middle-Light	Heavy-Heavy	Welter-Light Middle	Light-Light Welter	Bantam-Feather	Light Fly-Fly
Biacromiale	29.91**	Light Fly-Fly	Bantam-Feather	Light-Light Welter	Welter-Light Middle	Middle-Light	Heavy-Heavy
Biiliocristale	23.46**	Light Fly-Fly	Bantam-Feather	Light-Light Welter	Welter-Light Middle	Middle-Light	Heavy-Heavy
Arm length	15.28**	Light Fly-Fly	Bantam-Feather	Light-Light Welter	Welter-Light Middle	Middle-Light	Heavy-Heavy
Leg length	23.39**	Light Fly-Fly	Bantam-Feather	Light-Light Welter	Welter-Light Middle	Middle-Light	Heavy-Heavy
Trunk length	15.92**	Light Fly-Fly	Bantam-Feather	Light-Light Welter	Welter-Light Middle	Middle-Light	Heavy-Heavy

* $F_{.05}$ (4,134) = 2.44   ** $F_{.01}$ (4,134) = 3.46

TABLE 3.82. DESCRIPTIVE ANALYSIS FOR WEIGHTLIFTING, TOTAL GROUP

Variable	Units	N	Mean	S.D.	Max.	Min.
Age	yr.	59	26.7	6.1	49.0	17.0
Weight	kg.	59	76.6	20.4	141.0	52.9
Height, Max. stretch	cm.	59	168.0	11.4	200.8	147.2
Trunk length	cm.	59	52.3	4.0	64.0	45.4
Biacromial breadth	cm.	58	40.9	2.1	49.9	36.1
Biiliocristal breadth	cm.	59	28.0	2.8	35.6	23.7
Upper plus lower arm length, right	cm.	59	55.5	4.3	66.7	47.7
Total leg length,right	cm.	58	75.5	6.0	91.9	63.9
Sum 3 skinfolds	mm.	59	25.2	13.9	80.9	11.8
Height-Weight ratio		59	12.06	.42	12.73	10.85
Endomorphy		59	2.4	1.37	7.0	1.0
Mesomorphy		59	7.1	1.03	10.0	5.0
Ectomorphy		59	1.0	.48	2.0	.5

TABLE 3.83. DESCRIPTIVE ANALYSIS, WEIGHTLIFTING, BANTAM AND FEATHER WEIGHTS

Variable	Units	N	Mean	S.D.	Max.	Min.
Age	yr.	21	26.7	5.7	40.0	17.0
Weight	kg.	21	59.1	8.0	63.2	52.9
Height, Max. stretch	cm.	21	156.9	5.1	164.2	147.2
Trunk length	cm.	21	48.9	2.0	53.3	45.4
Biacromial breadth	cm.	21	38.4	1.3	40.9	36.1
Biiliocristal breadth	cm.	21	25.6	1.2	28.7	23.7
Upper plus lower arm length, right	cm.	21	51.4	2.1	56.4	47.7
Total leg length,right	cm.	21	69.6	3.2	74.7	63.9
Sum 3 skinfolds	mm.	21	17.0	2.5	20.8	11.9
Height-Weight ratio		21	12.20	.36	12.73	11.25
Endomorphy		21	1.5	.35	2.0	1.0
Mesomorphy		21	6.8	.91	9.0	5.0
Ectomorphy		21	1.1	.50	2.0	.5

TABLE 3.84. DESCRIPTIVE ANALYSIS FOR WEIGHTLIFTING, LIGHT AND MIDDLE WEIGHTS

Variable	Units	N	Mean	S.D.	Max.	Min.
Age	yr.	16	26.1	6.5	45.0	18.0
Weight	kg.	16	70.5	4.0	78.5	65.6
Height, Max. stretch	cm.	16	167.4	4.6	175.0	157.5
Trunk length	cm.	16	51.8	2.5	56.3	47.5
Biacromial breadth	cm.	15	40.9	1.3	43.8	36.5
Biiliocristal breadth	cm.	16	27.4	1.0	29.7	26.1
Upper plus lower arm length, right	cm.	16	55.5	2.5	60.5	51.7
Total leg length,right	cm.	16	75.9	2.6	79.7	70.5
Sum 3 skinfolds	mm.	16	22.3	11.9	65.0	11.8
Height-Weight ratio		16	12.27	.22	12.61	11.75
Endomorphy		16	2.1	1.11	6.0	1.0
Mesomorphy		16	6.8	.51	7.5	5.5
Ectomorphy		16	1.2	.41	2.0	.5

TABLE 3.85. DESCRIPTIVE ANALYSIS FOR WEIGHTLIFTING, LIGHT-HEAVY, MIDDLE-HEAVY AND HEAVY WEIGHTS

Variable	Units	N	Mean	S.D.	Max.	Min.
Age	yr.	22	27.2	6.4	49.0	18.0
Weight	kg.	22	97.8	18.1	78.1	62.9
Height, Max. stretch	cm.	22	178.9	8.5	200.8	167.7
Trunk length	cm.	22	55.9	3.2	64.0	50.9
Biacromial breadth	cm.	20	44.1	3.1	49.9	36.8
Biiliocristal breadth	cm.	22	30.8	2.4	35.6	27.2
Upper plus lower arm length, right	cm.	22	59.3	3.2	66.7	54.3
Total leg length,right	cm.	21	81.0	4.5	91.9	71.3
Sum 3 skinfolds	mm.	22	35.2	15.6	80.9	13.4
Height-Weight ratio		22	11.78	.44	12.40	10.85
Endomorphy		22	3.5	1.48	7.0	1.0
Mesomorphy		22	7.7	1.20	10.0	5.0
Ectomorphy		22	.7	.33	1.5	.5

TABLE 3.86. ANALYSIS OF DIFFERENCES BETWEEN GROUPS - WEIGHTLIFTING

Variable	F-ratio	Between group differences		
Height	68.16**	Bantam-Feather	Light-Middle	Heavy
Endomorphy	13.85**	Bantam-Feather	Light-Middle	Heavy
Mesomorphy	8.61**	Light-Middle	Bantam-Feather	Heavy
Ectomorphy	12.23**	Heavy	Bantam-Feather	Light-Middle
Biacromiale	15.00**	Bantam-Feather	Light-Middle	Heavy
Biiliocristale	48.11**	Bantam-Feather	Light-Middle	Heavy
Arm length	48.67**	Bantam-Feather	Light-Middle	Heavy
Leg length	55.67**	Bantam-Feather	Light-Middle	Heavy
Trunk length	39.01**	Bantam-Feather	Light-Middle	Heavy

* $F_{.05}$ (2,54) = 3.17 ** $F_{.01}$ (2.54) = 5.02

TABLE 3.87. DESCRIPTIVE ANALYSIS FOR REFERENCE GROUP, TOTAL MEN

Variable	Units	N	Mean	S.D.	Max.	Min.
Age	yr.	267	22.3	4.9	54.0	14.0
Weight	kg.	267	63.1	8.8	98.3	39.8
Height, Max. stretch	cm.	267	170.4	7.0	197.8	151.0
Trunk length	cm.	267	51.7	2.8	59.7	42.0
Briacromial breadth	cm.	267	39.1	1.9	44.5	33.4
Biiliocristal breadth	cm.	267	27.4	1.6	33.3	24.0
Upper plus lower arm length, right	cm.	266	57.2	2.9	67.3	49.1
Total leg length,right	cm.	266	78.7	4.4	94.3	64.9
Sum 3 skinfolds	mm.	266	33.5	14.0	87.7	13.6
Height-Weight ratio		266	12.99	.54	14.40	11.50
Endomorphy		265	3.3	1.37	7.5	1.0
Mesomorphy		265	4.6	.93	7.5	2.5
Ectomorphy		265	2.9	1.26	6.5	.5

TABLE 3.88. DESCRIPTIVE ANALYSIS FOR REFERENCE GROUP, TOTAL WOMEN

Variable	Units	N	Mean	S.D.	Max.	Min.
Age	yr.	96	19.1	3.5	29.0	12.0
Weight	kg.	88	53.9	7.6	82.0	34.0
Height, Max. stretch	cm.	88	158.2	6.4	177.2	141.0
Trunk length	cm.	88	48.4	2.2	54.9	43.3
Biacromial breadth	cm.	88	35.2	1.8	39.5	30.0
Biiliocristal breadth	cm.	88	27.2	2.1	34.6	19.1
Upper plus lower arm length, right	cm.	87	52.8	2.6	62.1	45.8
Total leg length, right	cm.	88	72.6	4.2	83.3	62.8
Sum 3 skinfolds	mm.	87	53.6	20.5	113.0	21.7
Height-Weight ratio		88	12.73	.58	13.90	11.10
Endomorphy		87	5.1	1.64	9.0	2.0
Mesomorphy		87	3.9	.99	7.5	2.5
Ectomorphy		87	2.3	1.19	5.0	.5

TABLE 3.89. ANALYSIS OF DIFFERENCES BETWEEN SELECTED SPORTS (MEN)

Variable	F-ratio	BETWEEN-SPORT DIFFERENCES[b]										
Height	47.43**	GYM	LDMT	DIV	PT	CYC	SPR	SWM	WTP	CAN	ROW	BKB
Weight	56.56**	LDMT	GYM	DIV	SPR	CYC	PT	SWM	CAN	WTP	BKB	ROW
Endomorphy	21.38**	GYM	LDMT	SPR	CYC	CAN	DIV	PT	BKB	ROW	SWIM	WTP
(Endomorphy)[a]	(8.55)**	(GYM	LDMT	SPR	CYC	CAN	DIV	PT	BKB	ROW	SWIM)	
Mesomorphy	14.70**	LDMT	BKB	CYC	SWM	SPR	WTP	PT	ROW	DIV	CAN	GYM
Ectomorphy	14.57**	WTP	ROW	PT	GYM	CAN	CYC	DIV	SPR	SWM	BKB	LDMT
Biacromiale	24.40**	LDMT	CYC	GYM	DIV	SPR	PT	SWM	CAN	WTP	ROW	BKB
Biiliac	25.86**	GYM	SPR	LDMT	DIV	SWM	CYC	PT	WTP	CAN	BKB	ROW
Arm length	31.79**	GYM	LDMT	DIV	CYC	PT	SPR	WTP	SWM	CAN	ROW	BKB
Trunk length	14.20**	GYM	LDMT	DIV	SPR	PT	SWM	CYC	CAN	WTP	ROW	BKB
Leg length	27.84**	GYM	DIV	LDMT	CYC	PT	SWM	WTP	SPR	CAN	ROW	BKB

* $F_{.05}$ (10,601) = 1.85   ** $F_{.01}$ (10,601) = 2.36

[a] Endomorphy (skinfold) analysis run without water polo group.

[b] A key to these abbreviations can be found in the caption to Figure 3.14.

TABLE 3.90. ANALYSIS OF DIFFERENCES BETWEEN SELECTED SPORTS (WOMEN)

Variable	F-ratio	Between-sport Differences[b]					
Height	8.61**	GYM	DIV	CAN	SWW	SPR	WGT
Weight	13.42**	GYM	DIV	SPR	SWW	CAN	WGT
(Weight)[a]	(5.53)**	(GYM	DIV	SPR	SWW	CAN)	
Endomorphy	11.12**	GYM	SPR	DIV	SWW	CAN	WGT
(Endomorphy)[a]	(1.87)	GYM	SPR	DIV	SWW	CAN	
Mesomorphy	5.81**	SPR	DIV	SWM	GYM	WGT	CAN
Ectomorphy	6.15**	WGT	CAN	GYM	DIV	SPR	SWM
Biacromiale	4.46**	GYM	DIV	SPR	SWM	CAN	WGT
Biiliac	8.94**	GYM	SPR	DIV	SWM	CAN	WGT
Arm length	5.39**	GYM	DIV	CAN	SWM	SPR	WGT
Trunk length	2.68**	GYM	DIV	SPR	CAN	SWM	WGT
Leg length	6.58**	GYM	DIV	CAN	SWM	SPR	WGT

* $F_{.05}$ (5,89) = 2.32 ** $F_{.01}$ (5,89) = 3.23

a Weight and endomorphy (skinfold) analyses run without the weight throwers.

b A key to these abbreviations can be found in the caption to Figure 3.14.

TABLE 3.91. NUMBER OF MALE SUBJECTS BY SPORT, EVENT AND RACE

Sport	Event	Caucasoid	Mongoloid	Mestizo	Negroid	Unknown	Total
Track	Sprint	18	5	7	49	2	81
	400 m	14		6	29		49
	Middle Distance	26		4	11		41
	Long Distance	16	3	10	5		34
	Marathon	7	3	7	3		20
	Walk	11	1	8	1		21
	TOTALS	92	12	42	98	2	246
Field	Jump	11	5	1	15		32
	Vault	5	1				6
	Weight Throw	11	1		4		16
	Javelin	4		1	3		8
	Decathlon	4	2	2	1		9
	TOTALS	35	9	4	23		71
Swimming	Free	15		6	1		22
	Breast	7	1	4			12
	Back	8	1	2			11
	Butterfly	6	1	4			11
	Medley	9	1	2			12
	TOTALS	45	4	18	1		68
Diving		14		2			16
Water Polo		58	1	8	3	1	71
Modern Pentathlon		17	3	4			24
Basketball		36	1	6	20		63
Rowing	Sculls	15		1			16
	Pairs	18		2			20
	Fours	19		4	1		24
	Eights	21		6			27
	TOTALS	73		13	1		87
Canoeing		39		8	2		49
Cycling	Sprint	8	3	2	2		15
	Pursuit	11	9	1			21
	Road	37	12	15	5		69
	TOTALS	56	24	18	7		105
Gymnastics		18	3	6	1		28
Wrestling	Free-Style	21	9	15	4	1	50
	Greco-Roman	34	1	4	1		40
	TOTALS	55	10	19	5	1	90
	Fly-Bantam	12	2	7	2		23
	Feather-Light	23	5	10	3	1	42
	Welter-Middle	14	2	2			18
	Heavy	6	1				7
Boxing	Light Fly-Fly	12	1	8	7		28
	Bantam-Feather	10	4	11	7		32
	Light-Light Welter	16	4	4	9		33
	Welter-Light Middle	11		8	10		29
	Middle-Heavy	6		5	8		19
	TOTALS	55	9	36	41		141
Weightlifting	Bantam-Feather	10	6	4	1		21
	Light-Middle	10	2	4			16
	Heavy	18		2	1		21
	TOTALS	38	8	10	2		58
All Athletes	TOTALS	631	84	194	204	4	1117
Reference Group		104		162	1		267

TABLE 3.92. NUMBER OF FEMALE SUBJECTS BY SPORT, EVENT AND RACE

Sport	Event	Caucasoid	Mongoloid	Mestizo	Negroid	Total
Track and field	Sprint	8	1	6	13	28
	Middle Distance	11		2	5	18
	Jump	5		1	6	12
	Throw	9		2	2	13
	Pentathlon	6		1	4	11
	TOTALS	39	1	12	30	82
Swimming		23	2	7		32
Diving		6		1		7
Canoeing		6				6
Gymnastics		13		7	1	21
All Athletes		87	3	27	31	148
Reference Group		45	1	49	1	96

TABLE 3.93. DESCRIPTIVE ANALYSIS BY RACE FOR MEN ATHLETES, CAUCASOID

Variable	Units	N	Mean	S.D.	Max.	Min.
Age	yr.	631	23.9	4.3	49.0	16.0
Weight	kg.	625	74.4	12.6	141.0	49.6
Height, Max. stretch	cm.	625	178.2	9.5	212.0	148.9
Trunk length	cm.	628	53.9	3.3	64.3	45.1
Biacromial breadth	cm.	628	41.2	2.5	49.9	28.5
Biiliocristal breadth	cm.	628	28.9	2.0	35.6	20.8
Upper plus lower arm length, right	cm.	623	59.9	3.8	74.1	47.7
Total leg length, right	cm.	619	82.1	6.1	106.0	63.5
Sum 3 skinfolds	mm.	626	21.4	8.4	80.9	11.2
Height-Weight ratio		625	12.85	.49	14.39	10.85
Endomorphy		624	2.0	.92	7.0	1.0
Mesomorphy		624	5.3	1.18	10.0	2.0
Ectomorphy		624	2.5	1.05	6.0	.5

TABLE 3.94. DESCRIPTIVE ANALYSIS BY RACE FOR MEN ATHLETES, MONGOLOID

Variable	Units	N	Mean	S.D.	Max.	Min.
Age	yr.	84	24.3	4.4	40.0	14.0
Weight	kg.	78	62.4	7.5	88.0	50.8
Height, Max. stretch	cm.	78	168.4	6.7	188.5	150.6
Trunk length	cm.	78	52.3	2.6	59.6	45.9
Biacromial breadth	cm.	78	39.3	1.9	44.2	35.2
Biiliocristal breadth	cm.	78	26.4	1.3	29.9	23.5
Upper plus lower arm length, right	cm.	78	55.8	2.8	64.6	48.5
Total leg length,right	cm.	78	76.3	4.5	91.1	66.7
Sum 3 skinfolds	mm.	78	18.8	3.9	31.4	11.9
Height-Weight ratio		78	12.87	.46	13.85	11.75
Endomorphy		78	1.7	.49	3.5	1.0
Mesomorphy		78	5.3	1.13	8.5	3.0
Ectomorphy		78	2.6	1.06	5.0	.5

TABLE 3.95. DESCRIPTIVE ANALYSIS BY RACE FOR MEN ATHLETES, MESTIZO

Variable	Units	N	Mean	S.D.	Max.	Min.
Age	yr.	194	23.4	4.4	40.0	15.0
Weight	kg.	194	65.9	8.5	89.8	47.5
Height, Max. stretch	cm.	194	170.9	7.7	189.1	154.3
Trunk length	cm.	193	52.1	2.9	59.3	45.6
Biacromial breadth	cm.	193	40.1	2.0	45.7	33.1
Biiliocristal breadth	cm.	192	27.8	1.6	32.0	23.9
Upper plus lower arm length, right	cm.	192	57.8	3.4	66.9	49.9
Total leg length,right	cm.	193	78.5	4.9	93.4	65.9
Sum 3 skinfolds	mm.	194	22.3	7.7	59.9	12.7
Height-Weight ratio		194	12.82	.39	13.95	11.76
Endomorphy		194	2.1	.86	6.0	1.0
Mesomorphy		194	5.3	.99	7.5	3.0
Ectomorphy		194	2.4	.89	5.0	.5

TABLE 3.96. DESCRIPTIVE ANALYSIS BY RACE FOR MEN ATHLETES, NEGROID

Variable	Units	N	Mean	S.D.	Max.	Min.
Age	yr.	204	24.1	4.5	39.0	16.0
Weight	kg.	203	69.1	9.8	106.8	47.5
Height, Max. stretch	cm.	203	176.8	7.9	200.9	147.2
Trunk length	cm.	201	51.9	2.7	61.4	45.3
Biacromial breadth	cm.	201	40.5	2.1	47.0	35.5
Biiliocristal breadth	cm.	201	26.5	1.5	30.9	22.7
Upper plus lower arm length, right	cm.	202	61.2	3.7	70.5	49.5
Total leg length,right	cm.	199	84.8	5.5	100.6	64.9
Sum 3 skinfolds	mm.	204	18.9	4.3	51.0	12.2
Height-Weight ratio		204	13.06	.44	14.20	11.25
Endomorphy		203	1.7	.54	5.0	1.0
Mesomorphy		203	4.8	.07	9.0	2.5
Ectomorphy		203	3.0	.03	6.0	0.5

TABLE 3.97. DESCRIPTIVE ANALYSIS BY RACE FOR WOMEN ATHLETES, CAUCASOID

Variable	Units	N	Mean	S.D.	Max.	Min.
Age	yr.	87	19.3	4.4	38.0	12.0
Weight	kg.	83	58.4	9.2	97.7	40.0
Height, Max. stretch	cm.	83	167.1	7.8	188.5	151.2
Trunk length	cm.	83	49.9	2.7	55.8	44.0
Biacromial breadth	cm.	83	37.3	1.8	42.6	33.3
Biiliocristal breadth	cm.	82	27.6	2.2	35.3	24.4
Upper plus lower arm length, right	cm.	82	55.3	3.2	63.8	48.6
Total leg length,right	cm.	81	77.7	4.8	95.0	68.7
Sum 3 skinfolds	mm.	83	29.3	12.7	91.4	14.1
Height-Weight ratio		83	13.08	.40	14.05	12.22
Endomorphy		83	2.9	1.25	8.0	1.0
Mesomorphy		83	3.9	.83	6.0	2.0
Ectomorphy		83	3.1	.97	5.5	1.0

TABLE 3.98. DESCRIPTIVE ANALYSIS BY RACE FOR WOMEN ATHLETES, MESTIZO

Variable	Units	N	Mean	S.D.	Max.	Min.
Age	yr.	27	17.8	3.6	28.0	14.0
Weight	kg.	26	54.5	10.6	83.7	43.5
Height, Max. stretch	cm.	26	159.6	7.4	176.2	150.0
Trunk length	cm.	26	48.5	3.1	56.6	43.3
Biacromial breadth	cm.	26	35.6	2.8	39.8	24.8
Biiliocristal breadth	cm.	27	26.5	2.0	32.3	22.9
Upper plus lower arm length,right	cm.	27	53.8	2.8	58.6	47.7
Total leg length,right	cm.	27	73.9	4.5	84.3	67.7
Sum 3 skinfolds	mm.	26	35.5	18.5	110.8	16.5
Height-Weight ratio		26	12.87	.46	13.67	11.31
Endomorphy		26	3.5	1.54	9.0	1.5
Mesomorphy		26	4.2	1.07	7.5	3.0
Ectomorphy		26	2.6	.87	4.5	.5

TABLE 3.99. DESCRIPTIVE ANALYSIS BY RACE FOR WOMEN ATHLETES, NEGROID

Variable	Units	N	Mean	S.D.	Max.	Min.
Age	yr.	31	21.3	3.5	29.0	15.0
Weight	kg.	31	55.7	8.3	89.3	45.6
Height, Max. stretch	cm.	31	164.9	6.6	178.5	151.1
Trunk length	cm.	31	47.9	2.9	55.5	42.0
Biacromial breadth	cm.	31	36.8	2.0	41.7	33.4
Biiliocristal breadth	cm.	31	25.5	1.7	31.1	23.2
Upper plus lower arm length, right	cm.	31	56.4	2.7	62.1	51.0
Total leg length,right	cm.	31	79.6	4.7	89.2	70.8
Sum 3 skinfolds	mm.	31	24.1	10.3	67.9	15.6
Height-Weight ratio		31	13.09	.43	14.00	12.05
Endomorphy		31	2.3	1.06	6.5	1.5
Mesomorphy		31	3.7	.80	5.5	2.0
Ectomorphy		31	3.1	1.05	5.5	1.0

TABLE 3.100. DESCRIPTIVE ANALYSIS BY RACE FOR REFERENCE MEN, CAUCASOID

Variable	Units	N	Mean	S.D.	Max.	Min.
Age	yr.	104	22.3	5.6	54.0	15.0
Weight	kg.	104	64.8	10.1	98.3	49.3
Height, Max. stretch	cm.	104	172.9	7.2	197.8	154.2
Trunk length	cm.	104	54.5	2.9	59.7	45.4
Biacromial breadth	cm.	104	39.0	2.0	44.5	34.2
Biiliocristal breadth	cm.	104	27.7	1.8	33.3	24.2
Upper plus lower arm length, right	cm.	104	57.9	3.1	67.3	50.0
Total leg length,right	cm.	104	79.9	4.5	94.3	70.2
Sum 3 skinfolds	mm.	103	33.5	14.9	76.8	13.6
Height-Weight ratio		103	13.07	.53	14.40	11.90
Endomorphy		102	3.3	1.46	7.0	1.0
Mesomorphy		102	4.5	.96	7.5	2.5
Ectomorphy		102	3.1	1.28	6.5	.5

TABLE 3.101. DESCRIPTIVE ANALYSIS BY RACE FOR REFERENCE MEN, MESTIZO

Variable	Units	N	Mean	S.D.	Max.	Min.
Age	yr.	162	22.2	4.5	42.0	14.0
Weight	kg.	162	62.0	7.7	84.0	39.8
Height, Max. stretch	cm.	162	168.7	6.4	185.7	151.0
Trunk length	cm.	162	51.3	2.7	57.8	42.0
Biacromial breadth	cm.	162	39.2	1.8	44.4	33.4
Biiliocristal breadth	cm.	162	27.3	1.5	32.1	24.0
Upper plus lower arm length, right	cm.	161	56.7	2.7	65.1	49.1
Total leg length,right	cm.	161	77.9	4.2	92.4	64.9
Sum 3 skinfolds	mm.	162	33.5	13.4	87.7	15.3
Height-Weight ratio		162	12.93	.53	14.23	11.50
Endomorphy		162	3.3	1.32	7.5	1.5
Mesomorphy		162	4.7	.90	7.0	2.5
Ectomorphy		162	2.7	1.22	6.0	.5

TABLE 3.102. DESCRIPTIVE ANALYSIS BY RACE, REFERENCE WOMEN, CAUCASOID

Variable	Units	N	Mean	S.D.	Max.	Min.
Age	yr.	45	19.3	3.2	29.0	12.0
Weight	kg.	41	53.6	7.0	73.5	40.4
Height, Max. stretch	cm.	41	161.6	6.0	177.2	150.4
Trunk length	cm.	41	48.9	2.1	53.3	43.3
Biacromial breadth	cm.	41	35.1	1.9	39.1	30.0
Biiliocristal breadth	cm.	41	27.7	1.7	31.9	24.8
Upper plus lower arm length,right	cm.	41	53.6	2.8	62.1	48.8
Total leg length,right	cm.	41	74.6	4.1	83.3	66.1
Sum 3 skinfolds	mm.	40	45.9	15.8	80.8	21.7
Height-Weight ratio		41	13.00	.40	13.75	12.32
Endomorphy		40	4.6	1.39	7.0	2.0
Mesomorphy		40	3.5	.70	5.0	2.5
Ectomorphy		40	2.9	.98	4.5	1.0

TABLE 3.103. DESCRIPTIVE ANALYSIS BY RACE FOR REFERENCE WOMEN, MESTIZO

Variable	Units	N	Mean	S.D.	Max.	Min.
Age	yr.	49	18.7	3.7	27.0	12.0
Weight	kg.	46	54.0	8.2	82.0	34.0
Height, Max. stretch	cm.	46	155.0	5.0	168.4	141.0
Trunk length	cm.	46	47.8	2.1	54.9	43.7
Biacromial breadth	cm.	46	35.2	1.6	39.5	31.4
Biiliocristal breadth	cm.	46	27.7	2.5	34.6	19.1
Upper plus lower arm length, right	cm.	45	52.1	2.3	56.3	45.8
Total leg length,right	cm.	46	70.7	3.4	79.3	62.8
Sum 3 skinfolds	mm.	46	60.7	21.9	113.0	24.6
Height-Weight ratio		46	12.48	.62	13.90	11.10
Endomorphy		46	5.7	1.67	9.0	2.5
Mesomorphy		46	4.3	1.04	7.5	2.5
Ectomorphy		46	1.8	1.17	5.0	.5

TABLE 3.104. ANALYSIS OF DIFFERENCES BY RACE IN SPRINTERS (MEN'S TRACK AND FIELD)

Variable	F-ratio	Mean	S	Mean	S	Mean	S	Mean	S
Height (cm)	2.82*	169.4 Mestizo(n=7)	6.5	173.5 Mongoloid(n=4)	7.9	175.9 Negroid(n=47)	5.2	176.7 Caucasoid(n=17)	7.7
Weight (kg)	1.73	63.9 Mestizo	4.8	66.1 Mongoloid	3.8	69.0 Caucasoid	6.0	69.5 Negroid	7.0
Skinfolds (mm) (Endomorphy)	1.61	18.4 Negroid	3.0	19.2 Caucasoid	3.0	20.4 Mongoloid	1.3	21.4 Mestizo	8.5
Mesomorphy	0.07	5.0 Mongoloid	.71	5.1 Negroid	.85	5.1 Caucasoid	1.2	5.2 Mestizo	.95
H. W. R. (Ectomorphy)	0.45	12.83 Mestizo	.38	12.96 Negroid	.37	12.98 Mongoloid	.33	13.02 Caucasoid	.40
Acromiale (cm)	1.59	38.9 Mongoloid	2.8	39.7 Mestizo	1.4	40.2 Negroid	1.8	40.9 Caucasoid	1.9
Biiliocristale(cm)	8.08**	26.4 Negroid	1.3	26.5 Mongoloid	1.1	27.1 Mestizo	0.8	28.0 Caucasoid	1.1
Arm length (cm)	4.39**	57.6 Mestizo	2.0	57.7 Mongoloid	2.0	59.5 Caucasoid	2.8	60.8 Negroid	2.8
Leg length (cm)	5.77**	78.0 Mestizo	4.2	79.5 Mongoloid	3.4	83.6 Caucasoid	4.6	84.3 Negroid	4.0
Trunk length (cm)	0.55	51.6 Caucasoid	2.6	51.7 Negroid	2.0	51.8 Mestizo	2.7	53.1 Mongoloid	2.0

* $F_{.05}(3,71) = 2.74$ ** $F_{.01}(3,71) = 4.08$

TABLE 3.105. ANALYSIS OF DIFFERENCES BY RACE IN 400 meters (men)

Variable	F-ratio	Mean	S	Mean	S	Mean	S
Height (cm)	3.37*	173.3 Mestizo(n=6)	2.0	178.0 Negroid(n=28)	5.0	180.3 Caucasoid(n=14)	7.4
Weight (kg)	0.70	65.3 Mestizo	3.8	68.3 Negroid	6.5	68.8 Caucasoid	6.9
Skinfolds (mm) (Endomorphy)	2.36	16.5 Caucasoid	2.3	16.7 Negroid	2.5	19.2 Mestizo	4.0
Mesomorphy	1.77	4.3 Caucasoid	.67	4.4 Negroid	.77	5.0 Mestizo	.84
H. W. R. (Ectomorphy)	1.51	13.02 Mestizo	.30	13.17 Negroid	.33	13.29 Caucasoid	.34
Acromiale (cm)	2.67	39.8 Negroid	1.5	40.4 Mestizo	1.7	40.9 Caucasoid	1.6
Biiliocristale(cm)	8.65**	26.5 Negroid	1.3	27.9 Mestizo	1.9	28.5 Caucasoid	1.7
Arm length (cm)	4.27*	58.5 Mestizo	1.7	60.4 Caucasoid	2.4	61.5 Negroid	2.5
Leg length (cm)	3.95*	80.8 Mestizo	1.6	84.4 Caucasoid	4.0	85.3 Negroid	3.6
Trunk length (cm)	1.47	51.7 Mestizo	1.8	52.3 Negroid	2.1	53.3 Caucasoid	2.5

* $F_{.05}(2,45) = 3.21$ ** $F_{.01}(2,45) = 5.12$

TABLE 3.106. ANALYSIS OF DIFFERENCES BY RACE IN MIDDLE DISTANCE RUNNERS (MEN)

Variable	Caucasoid Mean	(N=20) S	Negroid Mean	(N=11) S	F-ratio
Height (cm)	178.3	6.9	177.5	3.6	0.15
Weight (kg)	66.3	6.2	63.4	3.5	2.18
Skinfolds (mm) (Endomorphy)	16.8	2.1	16.8	1.2	.0001
Mesomorphy	4.3	0.63	4.0	0.57	2.84
H. W. R. (Ectomorphy)	13.30	0.33	13.46	0.24	2.23
Acromiale (cm)	40.2	1.8	39.9	1.4	0.19
Biiliocristale (cm)	28.4	1.9	26.0	1.3	15.46**
Arm length (cm)	59.5	3.3	62.3	1.3	7.01*
Leg length (cm)	83.5	5.0	86.9	2.6	4.64*
Trunk length (cm)	53.2	3.1	51.3	3.2	2.82

* $F_{.05}$ (1,35) = 4.12 ** $F_{.01}$ (1,35) = 7.42

TABLE 3.107. ANALYSIS OF DIFFERENCES BY RACE IN LONG DISTANCE AND MARATHON RUNNERS (MEN)

Variable	F-ratio	Mean	S	Mean	S	Mean	S	Mean	S
Height (cm)	9.06*	166.5 Mestizo(n=17)	6.6	167.3 Mongoloid(n=6)	3.1	172.2 Negroid(n=8)	2.8	174.4 Caucasoid(n=22)	4.6
Weight (kg)	2.15	56.4 Negroid	6.5	56.5 Mongoloid	2.2	57.8 Mestizo	5.9	60.5 Caucasoid	3.9
Skinfolds (mm) (Endomorphy)	12.99**	14.4 Caucasoid	2.1	15.0 Mongoloid	2.4	16.8 Negroid	3.8	20.2 Mestizo	3.6
Mesomorphy	5.54**	3.5 Negroid	.53	4.0 Caucasoid	.77	4.6 Mongoloid	.66	4.6 Mestizo	.64
H. W. R. (Ectomorphy)	7.44**	13.05 Mestizo	.34	13.13 Mongoloid	.29	13.43 Caucasoid	.33	13.60 Negroid	.29
Acromiale (cm)	0.73	38.4 Negroid	1.5	38.7 Mestizo	1.6	39.1 Mongoloid	1.6	39.2 Caucasoid	1.6
Biiliocristale(cm)	3.65*	26.2 Mongoloid	0.9	26.2 Negroid	1.5	27.1 Mestizo	1.3	27.5 Caucasoid	1.1
Arm length (cm)	7.68**	54.5 Mongoloid	1.4	56.9 Mestizo	2.5	58.6 Caucasoid	2.3	59.3 Negroid	1.4
Leg length (cm)	5.27**	77.1 Mestizo	4.8	77.1 Mongoloid	0.6	80.8 Caucasoid	3.1	81.4 Negroid	1.8
Trunk length (cm)	3.13*	49.9 Mestizo	2.0	50.4 Mongoloid	2.2	51.3 Negroid	1.3	51.8 Caucasoid	2.3

* $F_{.05}$ (3,49) = 2.80 ** $F_{.01}$ (3,49) = 4.22

TABLE 3.108. ANALYSIS OF DIFFERENCES BY RACE IN WALKERS

Variable	Caucasoid Mean	(N=11) S	Mestizo Mean	(N=7) S	F-ratio
Height (cm)	176.4	5.9	169.7	2.6	7.91*
Weight (kg)	65.6	4.0	59.5	3.8	10.25**
Skinfolds (mm) (Endomorphy)	17.5	1.3	18.7	1.8	2.77
Mesomorphy	4.6	0.63	4.6	0.67	.004
H. W. R. (Ectomorphy)	13.27	0.22	13.15	0.21	1.23
Acromiale (cm)	39.5	1.8	39.6	1.4	0.03
Biiliocristale (cm)	28.0	1.2	29.8	0.8	0.09
Arm length (cm)	59.2	2.8	57.8	1.8	1.35
Leg length (cm)	81.3	4.2	77.2	2.3	5.44*
Trunk length (cm)	54.5	2.1	52.8	1.0	3.83

* $F_{.05}(1,16) = 4.49$ ** $F_{.01}(1,16) = 8.53$

TABLE 3.109. ANALYSIS OF DIFFERENCES BY RACE IN JUMPERS (MEN)

Variable	F-ratio	Mean	S	Mean	S	Mean	S
Height (cm)	12.83**	172.3 Mongoloid(n=5)	1.2	182.3 Negroid(n=15)	5.8	188.7 Caucasoid(n=11)	7.4
Weight (kg)	12.22**	62.0 Mongoloid	3.1	74.3 Negroid	6.1	77.5 Caucasoid	6.4
Skinfolds (mm) (Endomorphy)	0.08	17.6 Mongoloid	2.8	18.0 Negroid	3.2	18.2 Caucasoid	3.2
Mesomorphy	2.04	4.0 Caucasoid	.82	4.6 Negroid	.75	4.7 Mongoloid	.45
H. W. R. (Ectomorphy)	2.34	13.10 Negroid	.29	13.17 Mongoloid	.15	13.35 Caucasoid	.33
Acromiale (cm)	3.80*	39.5 Mongoloid	1.8	41.5 Negroid	1.9	42.2 Caucasoid	1.7
Biiliocristale(cm)	17.04**	26.0 Mongoloid	1.5	26.8 Negroid	1.4	29.7 Caucasoid	1.5
Arm length (cm)	9.69**	56.9 Mongoloid	0.8	63.3 Negroid	3.0	63.7 Caucasoid	3.7
Leg length (cm)	8.86**	80.1 Mongoloid	2.1	88.6 Negroid	3.9	89.2 Caucasoid	5.3
Trunk length (cm)	6.06**	52.0 Mongoloid	0.7	52.8 Negroid	2.0	55.3 Caucasoid	2.6

* $F_{.05}(2,28) = 3.34$ ** $F_{.01}(2,28) = 5.45$

TABLE 3.110. ANALYSIS OF DIFFERENCES BY RACE IN SPRINTERS (WOMEN'S TRACK AND FIELD)

Variable	F-ratio	Mean	S	Mean	S	Mean	S
Height (cm)	10.35**	158.7 Mestizo(n=6)	4.0	164.2 Negroid(n=13)	4.6	170.5 Caucasoid(n=8)	5.7
Weight (kg)	3.19	52.4 Mestizo	5.5	56.1 Negroid	6.0	60.7 Caucasoid	6.8
Skinfolds (mm) (Endomorphy)	2.46	24.2 Negroid	8.6	27.7 Caucasoid	4.2	32.7 Mestizo	9.4
Mesomorphy	0.16	3.9 Caucasoid	.52	4.0 Negroid	.69	4.1 Mestizo	.86
H. W. R. (Ectomorphy)	1.47	12.86 Mestizo	.36	12.98 Negroid	.36	13.15 Caucasoid	.23
Acromiale (cm)	1.37	35.7 Mestizo	1.6	36.9 Negroid	1.9	37.1 Caucasoid	1.4
Biiliocristale(cm)	3.79*	25.3 Negroid	1.4	26.9 Mestizo	2.0	27.0 Caucasoid	1.2
Arm length (cm)	3.74*	52.8 Mestizo	2.4	55.4 Caucasoid	2.5	56.1 Negroid	2.4
Leg length (cm)	7.42**	72.5 Mestizo	3.7	78.8 Caucasoid	4.2	79.6 Negroid	3.6
Trunk length (cm)	8.14**	47.4 Negroid	2.4	48.3 Mestizo	1.3	51.5 Caucasoid	2.8

* $F_{.05}$ (2,24) = 3.40   ** $F_{.01}$ (2,24) = 5.61

TABLE 3.111. ANALYSIS OF DIFFERENCES BY RACE IN MIDDLE DISTANCE RUNNERS (WOMEN'S TRACK AND FIELD)

Variable	Caucasoid Mean	(N=11) S	Negroid Mean	(N=5) S	F-ratio
Height (cm)	168.0	2.9	162.5	9.2	3.36
Weight (kg)	55.4	3.6	50.2	3.9	7.33*
Skinfolds (mm) (Endomorphy)	21.5	5.2	18.2	1.1	1.84
Mesomorphy	3.4	0.66	3.2	0.76	0.31
H. W. R. (Ectomorphy)	13.34	0.22	13.35	0.47	0.006
Acromiale (cm)	36.0	1.2	35.5	1.0	6.01*
Biiliocristale (cm)	27.7	1.6	25.5	0.9	8.15*
Arm length (cm)	55.5	1.7	56.0	3.9	0.15
Leg length (cm)	78.8	2.1	78.6	6.4	0.01
Trunk length (cm)	49.6	1.8	47.1	1.7	6.60*

* $F_{.05}$ (1,14) = 4.60   ** $F_{.01}$ (1,14) = 8.86

TABLE 3.112. ANALYSIS OF DIFFERENCES BY RACE IN JUMPERS (WOMEN'S TRACK AND FIELD)

Variable	Caucasoid Mean	(N=5) S	Negroid Mean	(N=6) S	F-ratio
Height (cm)	173.7	9.2	166.5	5.8	2.51
Weight (kg)	58.4	5.5	54.6	2.9	2.21
Skinfolds (mm) (Endomorphy)	24.4	6.8	20.8	1.2	1.71
Mesomorphy	3.1	0.74	3.5	0.71	0.84
H. W. R. (Ectomorphy)	13.58	0.33	13.29	0.33	2.11
Acromiale (cm)	37.5	1.6	37.2	2.2	0.06
Biiliocristale (cm)	27.3	2.4	24.6	1.1	6.40*
Arm length (cm)	58.7	3.1	58.1	2.6	0.11
Leg length (cm)	83.8	7.0	81.7	5.2	0.32
Trunk length (cm)	49.1	1.8	46.9	3.3	1.89

* $F_{.05}$ (1,9) = 5.12 ** $F_{.01}$ (1,9) = 10.56

TABLE 3.113. ANALYSIS OF DIFFERENCES BY RACE IN SWIMMING (MEN)

Variable	Caucasoid Mean	(N=43) S	Mestizo Mean	(N=18) S	F-ratio
Height (cm)	180.0	6.7	177.7	4.7	1.79
Weight (kg)	73.4	6.9	69.9	5.7	3.58
Skinfolds (mm) (Endomorphy)	21.5	5.3	23.3	5.2	1.54
Mesomorphy	5.0	0.74	4.9	0.92	0.45
H. W. R. (Ectomorphy)	12.99	0.29	13.05	0.34	0.49
Acromiale (cm)	41.6	1.8	40.4	2.2	5.78*
Biiliocristale (cm)	28.2	1.6	27.7	1.1	1.54
Arm length (cm)	60.8	3.1	60.5	2.6	0.16
Leg length (cm)	82.1	4.0	82.0	3.0	0.02
Trunk length (cm)	54.7	2.9	53.9	2.1	1.08

* $F_{.05}$ (2,59) = 4.00 ** $F_{.01}$ (1,59) = 4.08

TABLE 3.114. ANALYSIS OF DIFFERENCES BY RACE IN SWIMMING (WOMEN)

Variable	Caucasoid Mean	(N=20) S	Mestizo Mean	(N=6) S	F-ratio
Height (cm)	166.0	6.8	159.9	7.6	3.55
Weight (kg)	57.3	8.6	56.7	13.0	0.02
Skinfolds (mm) (Endomorphy)	30.2	10.6	34.6[+]	6.7	0.78
Mesomorphy	4.0	0.78	4.2	0.52	0.24
H. W. R. (Ectomorphy)	13.07	0.40	12.94	0.29	0.59
Acromiale (cm)	37.4	1.7	36.1	1.7	2.96
Biiliocristale (cm)	27.4	2.2	26.3	1.3	1.14
Arm length (cm)	55.5	3.1	54.5	1.9	0.56
Leg length (cm)	76.7	3.9	75.5	6.2	0.32
Trunk length (cm)	50.0	2.9	47.6	1.6	3.59

* $F_{.05}$ (1,24) = 4.26 ** $F_{.01}$ (1,24) = 7.82 [+] N = 5

TABLE 3.115. ANALYSIS OF DIFFERENCES BY RACE IN BASKETBALL

Variable	Caucasoid Mean	(N=36) S	Negroid Mean	(N=20) S	F-ratio
Height (cm)	191.4	9.0	188.8	7.0	3.83
Weight (kg)	83.2	10.9	75.1	6.4	9.20**
Skinfolds (mm) (Endomorphy)	21.4	7.4	21.0	4.0	0.07
Mesomorphy	4.4	0.89	4.1	0.84	1.28
H. W. R. (Ectomorphy)	13.23	0.40	13.36	0.32	1.44
Acromiale (cm)	42.8	2.4	42.1	2.5	1.08
Biiliocristale (cm)	30.7	2.3	27.5	1.5	30.33**
Arm length (cm)	64.1	3.7	65.0	3.1	0.83
Leg length (cm)	89.2	5.9	90.6	5.5	0.80
Trunk length (cm)	57.7	2.7	54.4	1.9	23.51**

* $F_{.05}$ (1,54) = 4.02 ** $F_{.01}$ (1,54) = 7.12

TABLE 3.116. ANALYSIS OF DIFFERENCES BY RACE IN ROWING

Variable	Caucasoid Mean	(N=71) S	Mestizo Mean	(N=13) S	F-ratio
Height (cm)	186.3	5.5	178.7	3.8	22.6**
Weight (kg)	84.1	6.9	74.7	5.5	21.4**
Skinfolds (mm) (Endomorphy)	21.5	4.7	23.3	6.3	1.31
Mesomorphy	5.4	0.88	5.2	1.09	0.72
H. W. R. (Ectomorphy)	12.83	0.29	12.83	0.44	.0005
Acromiale (cm)	42.7	1.8	41.2	1.4	8.28**
Biiliocristale (cm)	30.1	2.0	29.6	1.9	0.57
Arm length (cm)	63.0	2.6	60.3	2.0	12.64**
Leg length (cm)	86.8	3.8	82.8	3.9	11.95**
Trunk length (cm)	55.6	2.9	54.9	1.7	0.71

* $F_{.05}$ (1,82) = 3.96 ** $F_{.01}$ (1,82) = 6.96

TABLE 3.117. ANALYSIS OF DIFFERENCES BY RACE IN CYCLING

Variable	F-ratio	Mean	S	Mean	S	Mean	S	Mean	S
Height (cm)	12.15**	170.1 Mongoloid(n=23)	3.9	172.3 Mestizo(n=18)	6.9	174.2 Negroid(n=7)	7.4	177.9 Caucasoid(n=53)	5.6
Weight (kg)	25.12**	61.0 Mongoloid	4.2	66.1 Mestizo	5.4	68.8 Negroid	7.7	73.2 Caucasoid	6.3
Skinfolds (mm) (Endomorphy)	0.42	18.7 Mongoloid	3.8	19.2 Caucasoid	4.2	19.5 Negroid	3.1	20.0 Mestizo	3.3
Mesomorphy	2.03	4.7 Mestizo	.52	4.7 Mongoloid	.85	5.0 Negroid	1.00	5.2 Caucasoid	.86
H. W. R. (Ectomorphy)	2.52	12.86 Caucasoid	.30	12.89 Negroid	.48	12.89 Mestizo	.28	13.07 Mongoloid	.33
Acromiale (cm)	3.90*	38.6 Mongoloid	1.8	39.7 Mestizo	2.4	40.3 Caucasoid	2.1	40.6 Negroid	1.2
Biiliocristale(cm)	29.91**	26.1 Negroid	1.0	26.5 Mongoloid	1.1	28.0 Mestizo	1.1	29.1 Caucasoid	1.4
Arm length (cm)	10.51**	36.5 Mongoloid	2.1	58.4 Mestizo	3.2	59.6 Caucasoid	2.4	61.6 Negroid	4.2
Leg length (cm)	10.61**	76.9 Mongoloid	3.2	78.5 Mestizo	4.2	81.4 Caucasoid	3.5	83.3 Negroid	5.0
Trunk length (cm)	4.50**	50.9 Negroid	2.9	53.2 Mestizo	2.5	53.2 Mongoloid	1.9	54.3 Caucasoid	2.5

* $F_{.05}$ (3.97) = 2.70 ** $F_{.01}$ (3.97) = 3.97

TABLE 3.118. ANALYSIS OF DIFFERENCES BY RACE IN MALE GYMNASTS

Variable	F-ratio	Mean	S	Mean	S	Mean	S
Height (cm)	1.98	162.1 Mongoloid(n=3)	3.8	166.1 Mestizo(n=5)	5.1	168.7 Caucasoid(n=18)	5.8
Weight (kg)	1.95	57.6 Mongoloid	6.0	59.8 Mestizo	2.1	62.9 Caucasoid	5.2
Skinfolds (mm) (Endomorphy)	0.23	16.1 Mestizo	2.4	16.4 Caucasoid	2.1	17.1 Mongoloid	0.6
Mesomorphy	0.24	5.7 Mongoloid	.58	5.9 Caucasoid	.95	6.1 Mestizo	.65
H. W. R. (Ectomorphy)	0.14	12.73 Mongoloid	.19	12.85 Caucasoid	.41	12.85 Mestizo	.27
Acromiale (cm)	0.10	39.5 Mongoloid	1.0	39.8 Mestizo	0.9	39.9 Caucasoid	1.7
Biiliocristale(cm)	4.88*	24.8 Mongoloid	1.3	26.5 Mestizo	1.0	26.9 Caucasoid	1.1
Arm length (cm)	2.12	53.0 Mongoloid	1.2	56.2 Mestizo	2.6	56.2 Caucasoid	2.7
Leg length (cm)	0.45	75.0 Mestizo	4.1	76.9 Caucasoid	4.3	78.4 Mongoloid	11.0
Trunk length (cm)	1.32	49.1 Mestizo	2.3	50.1 Mongoloid	1.0	51.1 Caucasoid	2.6

* $F_{.05}$ (2.23) = 3.42   ** $F_{.01}$ (2.23) = 5.66

TABLE 3.119. ANALYSIS OF DIFFERENCES BY RACE IN GYMNASTS (WOMEN)

Variable	Caucasoid Mean	(N=12) S	Mestizo Mean	(N=7) S	F-ratio
Height (cm)	159.0	4.6	152.4	2.1	12.46**
Weight (kg)	51.6	4.4	46.1	2.8	8.85**
Skinfolds (mm) (Endomorphy)	24.6	5.4	30.5	7.0	4.33
Mesomorphy	4.1	0.53	4.4	0.38	1.03
H. W. R. (Ectomorphy)	13.03	0.24	12.89	0.18	1.75
Acromiale (cm)	36.0	1.7	33.3	3.9	4.66*
Biiliocristale (cm)	26.0	1.3	24.8	0.8	4.02
Arm length (cm)	52.0	1.6	51.4	2.0	0.60
Leg length (cm)	73.1	2.8	70.5	2.4	4.09
Trunk length (cm)	48.3	1.7	46.1	2.7	4.53*

* $F_{.05}$ (1,17) = 4.45   ** $F_{.01}$ (1,17) = 8.40

TABLE 3.120. ANALYSIS OF DIFFERENCES BY RACE IN FREE-STYLE WRESTLERS (FLY THROUGH MIDDLE WEIGHTS)

Variable	F-ratio	Mean	S	Mean	S	Mean	S	Mean	S
Height (cm)	0.70	165.7 Mestizo(n=15)	7.0	166.8 Negroid(n=4)	1.3	167.4 Mongoloid(n=7)	7.2	169.4 Caucasoid(n=17)	8.0
Weight (kg)	1.77	62.7 Negroid	7.9	64.6 Mestizo	9.8	70.0 Mongoloid	10.0	71.5 Caucasoid	10.3
Skinfolds (mm) (Endomorphy)	1.46	18.2 Negroid	2.5	21.4 Caucasoid	6.3	22.3 Mongoloid	5.1	26.0 Mestizo	10.8
Mesomorphy	0.74	5.8 Negroid	1.26	6.2 Mestizo	.99	6.4 Caucasoid	1.17	6.6 Mongoloid	.99
H. W. R. (Ectomorphy)	1.41	12.33 Mongoloid	.32	12.38 Caucasoid	.39	12.53 Mestizo	.34	12.74 Negroid	.62
Acromiale (cm)	0.91	39.1 Negroid	1.9	39.3 Mestizo	1.9	40.0 Mongoloid	1.6	40.6 Caucasoid	3.1
Biiliocristale(cm)	3.76*	25.5 Negroid	1.2	27.3 Mongoloid	1.6	27.7 Mestizo	1.6	28.3 Caucasoid	1.6
Arm length (cm)	0.51	54.6 Mongoloid	4.0	54.9 Negroid	3.9	56.4 Mestizo	3.8	56.5 Caucasoid	4.2
Leg length (cm)	1.39	74.1 Mongoloid	4.8	76.2 Mestizo	5.2	77.7 Caucasoid	6.2	80.9 Negroid	6.8
Trunk length (cm)	3.40*	50.0 Negroid	1.1	50.4 Mestizo	3.0	53.0 Caucasoid	2.7	53.1 Mongoloid	3.3

* $F_{.05}$ (3.39) = 2.86   ** $F_{.01}$ (3.39) = 4.38

TABLE 3.121. ANALYSIS OF DIFFERENCES BY RACE IN BOXERS (LIGHT FLY THROUGH LIGHT WELTER WEIGHTS)

Variable	F-ratio	Mean	S	Mean	S	Mean	S	Mean	S
Height (cm)	2.33	163.4 Mestizo(n=22)	5.1	164.8 Mongoloid(n=9)	5.2	166.4 Caucasoid(n=37)	5.2	167.4 Negroid(n=23)	6.1
Weight (kg)	0.49	56.6 Mestizo	5.3	57.6 Negroid	5.7	58.2 Caucasoid	5.1	58.4 Mongoloid	3.8
Skinfolds (mm) (Endomorphy)	0.86	17.7 Mongoloid	2.1	18.1 Caucasoid	3.3	18.5 Negroid	2.5	19.3 Mestizo	3.7
Mesomorphy	2.19	4.8 Negroid	.86	5.1 Caucasoid	.74	5.3 Mestizo	.79	5.4 Mongoloid	.81
H. W. R. (Ectomorphy)	2.50	12.87 Mongoloid	.35	12.90 Mestizo	.29	13.02 Caucasoid	.30	13.14 Negroid	.41
Acromiale (cm)	.008	38.8 Caucasoid	2.2	38.8 Negroid	1.0	38.8 Mestizo	1.6	38.9 Mongoloid	1.4
Biiliocristale(cm)	6.66**	25.4 Negroid	1.3	26.5 Mongoloid	1.2	26.5 Mestizo	1.4	27.0 Caucasoid	1.3
Arm length (cm)	1.22	54.2 Caucasoid	9.6	54.9 Mestizo	3.3	56.0 Mongoloid	2.3	57.4 Negroid	2.6
Leg length (cm)	7.18**	74.1 Mongoloid	2.4	74.5 Mestizo	2.4	75.3 Caucasoid	3.9	78.9 Negroid	4.6
Trunk length (cm)	2.96*	49.5 Negroid	2.3	50.0 Mestizo	2.3	51.1 Caucasoid	2.5	51.6 Mongoloid	2.4

* $F_{.05}$ (3.87) = 2.72   ** $F_{.01}$ (3.87) = 4.03

TABLE 3.122. ANALYSIS OF DIFFERENCES BY RACE IN BOXING (WELTER AND LIGHT MIDDLE WEIGHTS)

Variable	F-ratio	Mean	S	Mean	S	Mean	S
Height (cm)	0.56	172.7 Caucasoid(n=11)	4.7	172.8 Mestizo(n=9)	4.7	175.0 Negroid(n=8)	6.1
Weight (kg)	1.15	68.7 Negroid	3.8	69.7 Caucasoid	2.4	70.8 Mestizo	2.4
Skinfolds (mm) (Endomorphy)	1.43	19.9 Negroid	3.7	20.8 Caucasoid	4.2	23.8 Mestizo	6.8
Mesomorphy	0.17	5.3 Negroid	1.06	5.4 Caucasoid	.74	5.6 Mestizo	.62
H. W. R. (Ectomorphy)	2.06	12.63 Mestizo	.29	12.70 Caucasoid	.32	12.92 Negroid	.34
Acromiale (cm)	0.27	40.8 Caucasoid	1.3	41.1 Mestizo	2.2	41.4 Negroid	1.9
Biiliocristale(cm)	17.54**	25.7 Negroid	1.1	27.9 Mestizo	1.3	28.7 Caucasoid	1.1
Arm length (cm)	1.23	57.4 Mestizo	2.2	58.7 Caucasoid	1.9	59.5 Negroid	3.9
Leg length (cm)	0.96	79.3 Caucasoid	3.0	80.3 Mestizo	5.3	82.4 Negroid	6.7
Trunk length (cm)	0.57	52.2 Caucasoid	2.0	52.5 Negroid	2.0	53.2 Mestizo	2.4

* $F_{.05}\ (2,25) = 3.38$ ** $F_{.01}\ (2,25) = 5.57$

TABLE 3.123. ANALYSIS OF DIFFERENCES BY RACE IN WEIGHT LIFTERS (BANTAM THROUGH MIDDLE WEIGHTS)

Variable	F-ratio	Mean	S	Mean	S	Mean	S
Height (cm)	1.00	158.9 Mongoloid(n=8)	6.1	162.5 Mestizo(n=8)	5.8	162.8 Caucasoid(n=20)	7.4
Weight (kg)	1.76	60.4 Mongoloid	5.8	64.2 Mestizo	5.8	65.5 Caucasoid	7.1
Skinfolds (mm) (Endomorphy)	0.15	17.9 Mongoloid	2.2	19.7 Mestizo	4.1	19.7 Caucasoid	11.1
Mesomorphy	0.62	6.7 Mongoloid	.80	6.7 Caucasoid	.70	7.0 Mestizo	.46
H. W. R. (Ectomorphy)	0.18	12.23 Caucasoid	.26	12.28 Mongoloid	.36	12.29 Mestizo	.18
Acromiale (cm)	0.20	38.8 Mestizo	1.9	39.2 Mongoloid	1.2	39.4 Caucasoid	2.9
Biiliocristale(cm)	3.91*	25.5 Mongoloid	1.3	26.3 Mestizo	0.6	26.9 Caucasoid	1.5
Arm length (cm)	0.12	52.9 Mongoloid	2.3	53.1 Mestizo	2.1	53.5 Caucasoid	3.6
Leg length (cm)	1.59	70.3 Mongoloid	4.1	73.1 Caucasoid	4.0	73.4 Mestizo	4.6
Trunk length (cm)	0.02	50.2 Caucasoid	2.8	50.3 Mestizo	1.8	50.4 Mongoloid	3.2

* $F_{.05}\ (2,33) = 3.29$ ** $F_{.01}\ (2,33) = 5.32$

TABLE 3.124. ANALYSIS OF DIFFERENCES BY RACE IN THE REFERENCE GROUP (MEN)

Variable	Caucasoid Mean	(N=102) S	Mestizo Mean	(N=160) S	F-ratio
Height (cm)	172.9	7.1	168.9	6.4	23.42**
Weight (kg)	64.7	9.6	62.1	7.8	5.62*
Skinfolds (mm) (Endomorphy)	33.6	15.0	33.5	13.4	0.003
Mesomorphy	4.5	0.96	4.8	0.91	5.67*
H. W. R. (Ectomorphy)	13.08	0.54	12.93	0.55	4.55*
Acromiale (cm)	39.1	1.9	39.2	1.8	0.39
Biiliocristale (cm)	27.7	1.8	27.3	1.5	2.73
Arm length (cm)	57.9	3.0	46.7	2.7	10.91**
Leg length (cm)	79.9	4.5	77.9	4.3	12.02**
Trunk length (cm)	52.5	2.8	51.3	2.7	10.78**

* $F_{.05}$ (1,260) = 3.88 ** $F_{.01}$ (1,260) = 6.76

TABLE 3.125. ANALYSIS OF DIFFERENCES BY RACE IN THE REFERENCE GROUP (WOMEN)

Variable	Caucasoid Mean	(N=40) S	Mestizo Mean	(N=46) S	F-ratio
Height (cm)	161.4	5.9	155.0	5.0	29.63**
Weight (kg)	53.5	7.1	54.0	8.2	0.10
Skinfolds (mm) (Endomorphy)	45.9	15.8	60.7	21.9	12.67**
Mesomorphy	3.5	0.70	4.4	1.04	20.85**
H. W. R. (Ectomorphy)	13.00	0.40	12.48	0.62	20.86**
Acromiale (cm)	35.1	2.0	35.2	1.6	0.09
Biiliocristale (cm)	27.7	1.7	27.7	2.5	0.009
Arm length (cm)	53.5	2.8	52.1	2.3	6.67*
Leg length (cm)	74.4	4.0	70.7	3.5	20.98**
Trunk length (cm)	48.9	2.1	47.8	2.2	5.22**

* $F_{.05}$ (1,84) = 3.96 ** $F_{.01}$ (1,84) = 6.96

4

Family Studies

I. Family Studies of Ability

A. *Introductory Remarks*

A number of physiological and pathological characteristics in man are known to be partially influenced by heredity. It is further known that certain special talents (e.g., musical, mathematical, linguistic) are also partially hereditary. Studies of the inheritance of human abilities have a long history in modern times. Perhaps the first systematic studies of this type were those of Francis Galton (1885, 1889) while the record of a recent symposium on this subject is provided by Meade and Parkes (1966). Thus, it would be interesting to see whether certain physical abilities, such as those exhibited in sport events, are also to a certain extent conditioned by heredity.

This is the underlying reason that we decided to make an attempt, however imperfect, in this direction. Any group of outstanding athletes would provide us with a starting point for a study of this nature. The Olympic athletes are, of course, the best *propositi* any investigator can hope to have.

The study of possible genetic components in athletic ability is not easy. In studying the heritability of any trait we need information on the family members with respect to that trait. Even when the trait is a well-defined clinical entity, there are still many practical difficulties involved in tracing the pedigrees of the index cases, whether this is done by a clinician, an interviewer, or a questionnaire.

The two major difficulties in our present family study are (*1*) the lack of constant objective criteria from generation to generation, for measurement of sport ability, and (*2*) the absence of immediate family members of athletes in Mexico City during the Olympic Games period. In spite of this, the present study represents a pioneer effort in the definition of the hereditary and environmental factors which go to make up a successful athlete. Techniques used previously in the study of the inheritance of other aspects of human ability are appropriate to some extent but there is need also for new methodological approaches which these and subsequent similar studies may be expected to produce.

B. *Scale and Family Information*

With respect to the first difficulty—the lack of a quantitative measure of sport ability—we substitute a rough scale of 0, 1, 2, 3; where 0 indicates no sport ability at all, 1 indicates some sport ability but not significant enough to be considered an athlete, 2 indicates better than average sport ability, good enough to take part in local athletic events, and 3 indicates very high ability. On this scale, not only all Olympic athletes but all those on college, provincial, national, and professional teams belong to class 3. This scale, being rough and subjective, is in principle no different from those employed in many other fields of research in human behavior.

With respect to the second difficulty—the lack of direct information on immediate family members of the athletes—we used a simple questionnaire and an interviewer who asked the athlete to evaluate as objectively as he could the sport ability of his father, mother, and siblings. The information so obtained obviously depends on the responsivity of the athlete. In this respect some interesting patterns emerged out of the results which will be presented subsequently.

In all surveys intended to evaluate the role of heredity in the determination of a human trait, certain basic information about the family is useful. This is so whether the trait is one controlled by a single gene or whether it is a more complex one such as athletic ability determined by multiple hereditary and environmental factors. Much of this information is relatively easy to obtain, and in the questionnaire we asked for data concerning the longevity of parents, their age at the birth of the athlete, whether the athlete had a twin, and the athlete's order of birth. Questions have also been included concerning the geographical origin of the parents and grandparents of the athletes, and concerning consanguinity between the parents. Although consanguinity between parents is more usually of importance in the study of rare recessive traits, Darlington (1960) has argued that inbreeding may in certain cases have a favorable effect on human abilities. On the other hand, extreme outbreeding has resulted in the production of hybrid vigor in plants, though the study of Morton, Chung and Mi (1967) in Hawaii raises doubts whether a phenomenon of this type occurs in man.

C. *Reference Group*

For comparison about 370 random individuals in Mexico City were interviewed by the same interviewers using the same questionnaire. About 70% of these individuals were students of preparatory schools (beyond high school and roughly equivalent to American junior colleges), the rest being university students and Mexican government employees. These individuals were between ages 14 and 30, in the same general range of age as the athletes studied.

D. *Results*

1. *Parents*

The results of the evaluation of the parents of athletes are shown in Tables 4.1 (father), 4.2 (mother), and 4.3 (father and mother). These are pooled results from both sexes of all races and all sport specialities. Although the data may be easily sorted according to sex, race, and sports, we will not present the details keeping in mind our limited purposes and the nature of the data.

If either family-environment or genetic factors influence the sport ability of an individual, we should expect that proportionally more parents of the Olympic athletes have shown athletic ability than those of the control group. However, Table 4.1 shows that while in the control group, 50% of the fathers are reported to have no athletic ability at all, the Olympic athletes reported that 56% of their fathers belong to Class 0; this is clearly incompatible with either the environmental or genetic theories of sport ability but readily understandable in terms of human judgment. It was apparent that the highly accomplished athletes had a tendency to underestimate the sport ability of their parents (and others in general).

TABLE 4.1. SPORTS ABILITY OF ATHLETE'S FATHER

	0	1	2	3	Total
Athlete					
N	686	274	183	78	1,221
%	56.16	22.44	14.98	6.38	99.96
Control					
N	185	91	75	15	366
%	50.54	24.86	20.49	4.09	99.98

TABLE 4.2. SPORTS ABILITY OF ATHLETE'S MOTHER

	0	1	2	3	Total
Athlete					
N	1025	118	66	23	1,232
%	83.19	9.57	5.35	1.86	99.97
Control					
N	275	63	21	8	367
%	74.93	17.16	5.72	2.17	99.98

TABLE 4.3. MID PARENT (SPORTS ABILITY)

	0	1	2	3	4	5	6
Athletes							
N	646	221	184	92	44	14	8
%	53.43	18.27	15.21	7.60	3.63	1.15	0.66
Control							
N	151	92	71	36	11	5	2
%	41.03	25.00	19.29	9.78	2.98	1.35	0.54

TABLE 4.4. SPORTS ABILITY OF FATHER

	1	2	3	Total
Athletes				
N	273	183	78	534
%	51.12	34.27	14.61	100
Controls				
N	91	75	15	181
%	50.28	41.43	8.29	100

That the tendency to underestimate the athletic ability of others on the part of Olympic athletes is a real one may be illustrated by an example. One athlete has rated his younger brother who is also an Olympic delegate as belonging to Class 2, not Class 3. In his eyes, his brother is only better than average and not quite of the highest caliber. (Our more objective interviewer put his brother in Class 3.)

In spite of their tendency to underestimate others, the Olympic athletes reported more (6%) fathers with very high athletic ability (Class 3, Table 4.1) than the controls (4%). Although the tendency to underestimate parents appears to be a general one, probably affecting all classes, it seems that the greatest effect is to be found in Class 0. If that class is omitted as in Table 4.4, we would have a somewhat different picture.

The abridged table shows that there exists some association between fathers' sports ability and that of their children. The chi-square value is 6.09 for two degrees of freedom ($0.05 > p > 0.01$). Proceeding further in the same way, we may omit Class 1 from Table 4.4 to eliminate more bias due to underestimation. If we do so, the resulting table for Classes 2 and 3 yields a chi-square value of 6.01 for one degree of freedom ($0.05 > p > 0.01$).

Table 4.3 gives the results of the total score of father and mother for the two samples. If both father and mother are classified as 0, their combined score is 0. If one parent is 1 and the other 0, their combined score is 1. When the score is 3 it includes four situations, viz., $3 + 0$; $0 + 3$; $1 + 2$ and $2 + 1$. When the combined score is 6, it implies that both parents have been classified as 3. The total score, on being divided by 2, yields what is usually known as the "midparent" score. The underestimation effect is, of course, more pronounced in Table 4.3 than in the previous two tables. Thus 53% of the athletes report that both their fathers and mothers have no sport ability at all, while only 41% of the control individuals report so.

Physical abilities (e.g. running, jumping, swimming) are part of man's natural endowment analogous to musical and mathematical abilities, but the participation in arbitrarily defined competitive events under artificial regulations—which is sport—is a cultural manifestation of Western civilization. In other words, sport and games, being highly ritualized, are as much cultural events as physical exercises. People who have not been previously exposed to the sports culture of the West cannot be expected to do as well in competition as Westerners, even if they possess equal physical capabilities, the history of sports and games being much longer in Western industrialized countries than in others. This disparity was even greater in previous generations than in the present one. Therefore there is probably for the athletes from the developing countries no fair way for evaluation of parental athletic ability.

II. Order of Birth of Athletes

Altus (1966) reviewed the evidence that shows the ordinal position among siblings to be related to potential eminence and educational attainment. Most of the evidence seems to show that the eldest son has a greater likelihood of achieving eminence, having higher education, and having higher intelligence. "The dice are loaded in favor of the first-born." There is also some evidence to show that the youngest child is the next most favored one. Furthermore, Huntington (1938) argued that the first-born probably tend to be physically stronger and healthier than the younger siblings. The consensus is that the birth-order effect is due to social factors rather than biological ones. Naturally we were curious to see whether there is also an excess of first-borns among the Olympic athletes, as is the case with academic achievements.

Of the 1229 athletes on whom the information is available, 105 are from only-child families and are omitted from the study. The order of birth of the remaining 1124 athletes is shown in Table 4.5. By family size we really mean the size of sibship, of which the athlete is a member. Let s be the size of sibship and N_s be the number of sibships of size s. For example, $s = 5$, $N_5 = 128$ in Table 4.5. Each athlete is from a

TABLE 4.5. ORDER OF BIRTH OF THE ATHLETE

Family size	Family number	1	2	3	4	5	6	7	8	9	10	11	14
2	246	104	142										
3	233	82	88	63									
4	193	41	68	43	41								
5	128	20	31	33	25	19							
6	86	7	25	16	12	19	7						
7	81	14	11	19	8	15	8	6					
8	53	6	11	8	12	3	5	2	6				
9	39	8	6	2	2	5	2	4	5	5			
10	29	2	5	3	3	6	3	2	2	1	2		
11	14	1	2	2	1	2	0	2	1	0	1	2	
12	12	2	1	0	0	4	2	0	1	0	2	0	
13	7	0	3	0	1	1	0	1	1	0	0	0	
14	2	0	0	0	1	0	0	0	0	0	0	0	1
15	1	0	0	1	0	0	0	0	0	0	0	0	0
Total ($s \geq 2$)		287	393										
Total ($s \geq 3$)		183	251	190		Total last births 152							

different family; there are no brothers or sisters among the 1124 athletes in Table 4.5. Before proceeding to specific tests, we may mention a general method of calculating the chi-square. If Y is the observed number, EY is the expected number, and $V(Y)$ is the variance of Y, then

$$\chi^2 = \frac{(|Y - EY| - \frac{1}{2})^2}{V(Y)}$$

with a single degree of freedom. The one percent significance level requires the chi-square (with 1 df) to be greater than 6.63. The subsequent analysis is based on Mantel and Halperin (1963).

The most outstanding and surprising feature of Table 4.5 is that the second birth, not the first, seems to be in excess among the athletes. This will be our first test. The observed total number of second births is $Y = 393$. The expected number is $EY = 246/2 + 233/3 + \cdots = 317.30$. For any fixed sibship size $s \geq 2$, the probability of being a second birth is $1/s$ and not being a second is $(s - 1)/s$. The variance of the number of second births is thus

$$N_s\left(\frac{1}{s}\right)\left(\frac{s-1}{s}\right).$$

The sum of these quantities for all values of s is $V(Y) = 206.83$, the variance of the grand total number of second births for families of all sizes. The chi-square with one degree of freedom is $\chi^2 = 27.34$. The overall excess of the second births is statistically significant. These calculations have been summarized in Table 4.6.

TABLE 4.6. CHI SQUARE TESTS FOR ORDER OF BIRTHS

Test	Observed	Expected	Variance	Chi square
2nd births	393	$\Sigma \frac{N_s}{s}$ = 317.30	$\Sigma \frac{N_s(s-1)}{s^2}$ = 206.83	27.34
1st+last ($s \geq 3$)	183+152 = 335	$\Sigma \frac{2N_s}{s}$ = 388.60	$\Sigma \frac{2N_s(s-2)}{s^2}$ = 192.73	14.63
2nd vs 1st	393	1/2(287+393) = 340.0	1/2x1/2(680) = 170.00	16.21
2nd vs 3rd	251	1/2(251+190) = 220.5	1/2x1/2(441) = 110.25	8.16
2nd vs(1st+3rd)	393	1/2(246)+1/3(624) = 331	1/4(246)+2/9(624) = 200.17	18.90

In studies of order of births, much attention has been focused on the first and last births. Our second test, somewhat opposite to the first, was to see whether the combined frequency of the first- and the last-born is in deficit of expected. To do this we omit the 246 families of size 2, in which the last confounds with the second. Starting with families of size 3, we find that the total first-born are 183 and the total last-born are $63 + 41 + \cdots = 152$, so that the combined observed number is 335. The expected combined number is the sum of quantities

$$\left(\frac{2}{s}\right) N_s.$$

This turns out to be 388.60. It checks with $2(317.30) - 246 = 388.60$ correctly. For any fixed $s \geq 3$, the variance of the number of (first + last) births is

$$N_s \left(\frac{2}{s}\right)\left(\frac{s-2}{s}\right).$$

The sum of such quantities over s is the variance of the grand total number of (first + last) births. These calculations have been summarized in Table 4.6. The chi-square is 14.63 with one degree of freedom and, again, it is statistically significant.

However, as pointed out by Mantel and Halperin (1963), the data on birth order, especially those on diseases with a certain age of onset, are often biased, showing an excess of early born and a deficit of late born, simply because the latter group, being on the average younger, have not yet all reached the age of onset for the disease. Since one has to be of a certain minimum age before he is able to be an athlete, we wonder whether a similar bias exists in our data. Most of the athletes are young and their younger siblings may not have reached the age of athletes. On the other hand, their older siblings may be too old to be competing in Olympic games. The situation is not as clearcut as the case with disease of a certain age of onset. Should a bias exist, then it would be more appropiate to analyze the data on siblings close in age rather than on the entire data with very wide range in age. It is for this reason that we have performed the following three additional tests.

(*1*) Using the first two columns of Table 4.5 only, we wish to test whether the number of second births is greater than that of the first. (*2*) Starting from families of size 3 and using the second and third columns of Table 4.5 only, we test whether the number of second births is greater than that of the third. (*3*) Finally, using the first three columns of Table 4.5 we test the number of second births against the combined number of first and third births. These calculations have been summarized in the lower portion of Table 4.6. It is seen that in each case the difference is significant as measured by the chi-square test. Our general conclusion is that there is an excess of second births among the Olympic athletes.

The results reported earlier, needless to say, await the confirmation or refutation of later investigations. This being the case, we shall not attempt to advance any explanations at this stage. After a thorough review of the birth-order effects and social parameters involved, Altus (1966) concludes: "Ordinal position at birth has been shown to be related to significant social parameters, though the reasons behind the relations are as yet unknown or at best dimly apprehended." If it is so for social traits, it may even be more so for a trait like sports ability.

Bibliography

Altus, W. D. (1966). Birth order and its sequelae. *Science* **151**, 44–49.

Darlington, C. D. (1960). Cousin marriage and the evolution of the breeding system in man. *Heredity* **14**, 297.

Falconer, D. S. (1960). "Introduction to Quantitative Genetics." Oliver and Boyd, Edinburgh.

Galton, F. (1885). "Inquiries into Human Faculty and its Development." Macmillan, London.

Galton, F. (1889). "Natural Inheritance." Macmillan, London.

Huntington, E. (1938). "Season of Birth." Wiley, New York.

Malécot, G. (1962). Migration et parenté génétique moyenne. *In* "Les Déplacements Humains" (J. Sutter, ed.). Hachette, Paris.

Mantel, N., and Halperin, M. (1963). Analyses of birth-rank data. *Biometrics* **19**, 324–340.

Meade, J. E., and Parkes, A. S. (Ed.) (1966). "Genetic and Environmental Factors in Human Ability. Oliver and Boyd, Edinburgh.

Morton, N. E., and Yasuda, N. (1962). The genetical structure of human populations. *In* "Les Déplacements Humains" (J. Sutter, ed.). Hachette, Paris.

Morton, N. E., Chung, C. W., and Mi, M. P. (1967). Genetics and interracial crosses in Hawaii. "Monographs in Human Genetics" (vol. 3) Karger, Basel.

Penrose, L. S. (1939). Maternal age, order of birth and developmental abnormalities. *J. Ment. Sci.* **85**, 1141.

Penrose, L. S. (1955). Parental age and mutation. *Lancet* (ii), 312.

Penrose, L. S. (1961). Mutation. *In* "Recent Advances in Human Genetics" (L. S. Penrose, ed.). Churchill, London.

5

Taste Thresholds for Phenylthiourea (PTC)

I. Introduction

Among the reasons for including the PTC polymorphism among the genetic traits which might have a bearing on athletic performance were the following:

1. Tasters and nontasters though both normal nevertheless react differently to challenges to the thyroid so that, for instance, different types of goiter are developed by each of the types (Harris *et al.*, 1949; Kalmus, 1962). Thyroid function is certainly concerned with energy release in the body, and endocrine involvement must play an important role in meeting the stresses and challenges of competitive athletics.
2. Tasting and nontasting are the nearest one gets to a monofactorial genetic difference in normal human behavior.
3. The nontaster frequency varies greatly between human populations though the nontaster gene is probably present in most of them, excepting perhaps small isolated tribes.

4. The technique by Harris and Kalmus (1949) has now been applied to many populations which can serve as control populations for the present investigations.
5. This technique can be easily taught to comparatively unskilled people whose proficiency can be cross-checked without too much effort or time.

The present investigation was justified by the discovery in the data of what appears to be a significant negative association between nontasting and several categories of sport.

A. *The Taster Polymorphism*

In 1932 A. L. Fox synthesized a substance, ethoxyphenylthiourea, which proved to be intensely bitter to some people and tasteless to others. Subsequently a whole family of substances with this curious property were found, all of which contained a specific chemical group, the —NC═S group. Many of these occur in wild plants and vegetables and they number among them one class of antithyroid drugs, the thiouracils. The substance most widely used by geneticists is phenylthiourea (phenylthiocarbamide, PTC), which has also been used in this study.

When larger population samples were tested with serial dilutions of PTC it emerged that the difference between "tasters" and "nontasters" was not absolute, but that the threshold concentrations were distributed in a bimodal fashion. This situation obtains in all Caucasoid populations and many others with the exception of remote peoples—for instance, in Africa and Latin America. Figure 5.1 shows the threshold distribution for one of our control populations derived from urban Mexicans.

The difference in taste sensitivity was soon seized upon by geneticists and it emerged after a short time that "tasting" (that is, high sensitivity to PTC) was a single autosomal dominant character, and "nontasting" (that is, insensitivity to PTC) the recessive allelic character. We thus have two phenotypes, namely: *tasters,* which are either homozygous *TT* or heterozygous *Tt*; and *nontasters,* which are homozygous *tt.*

Family studies have shown that nontasters breed true while matings involving tasters produce offspring segregating in the expected frequencies.

Actually as a group the homozygous tasters are slightly more sensitive to PTC than the heterozygotes, but it is impossible to decide to which group an isolated individual belongs.

While the main difference between tasters and nontasters is determined by the allelic difference described above, the minor variation in sensitivity around the taster mode and the nontaster mode are probably caused by other genes as well as such factors as age, sex, and environment (Kalmus, 1965).

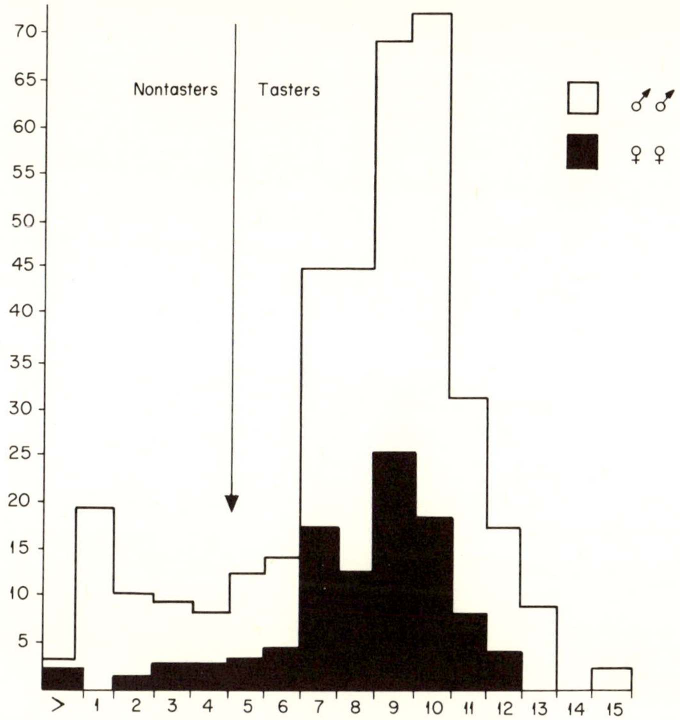

Figure 5.1 Taste threshold distribution for 365 urban Mexicans. Solution #1 contains 1.3 gm phenylthiourea/liter, solution #2 half this amount, solution #3 one quarter, and so on. The frequency of nontaster (0.134 ± 0.01) is intermediate between that for European populations, which is higher, and for tribal Americans, which can be nil.

II. Methods

Several months before the Games, a group of young scientists were trained in taste-testing following the method of Harris and Kalmus (1949) and their results were assessed by cross-testing and repetition. Details of the procedure can be read up in the original and only a summary will be given here.

A. Utensils and Solutions

Solution #1 was made up by dissolving 2.6 grams of phenylthiourea (BDH) in two liters of hot water. This was filtered and 1 liter poured into bottle #1. The other liter was diluted with 1 liter of water of which again one half was poured into bottle #2 and the other similarly diluted to give solution #3, and so on to solution #13.

Several series of 1-liter bottles labeled W (for water) were also available. These were filled with electropurified water, the same as was used for the solutions.

The solutions kept easily for the duration of the tests. Care was taken to ensure that all liquids used at a time had the same (room) temperature because people can distinguish between sips differing in temperature by less than 1°C.

Five to ten milliliters of the liquids were poured into disposable paper cups which were presented in metal holders. For each test eight cups and holders were used, four of which were marked as containing the solution while the others were used for the water. The marking was concealed from the athletes.

B. *Test Procedure*

Every test was done in two stages: In the preliminary test one beaker containing solution #9 was offered to the athlete who was then asked whether it had any taste and what it was. If the answer was that it was bitter or that it definitely was not water a weaker solution, for instance, #10 was offered until the taste was said to have disappeared. In each case the lowest concentration (highest solution number) at which a positive sensation was reported was taken as the *provisional* threshold.

The *definite* threshold[1] was ascertained by the "*eight tumbler method*," which is based on the celebrated tea-tasting experiment by R. A. Fisher. The athlete was presented with a set of eight cups, four of which contained 5 to 10 milliliters of his provisional threshold solution while the other four contained water. The athlete was asked to look away while the cups were randomly arranged on the table. He was than told that four of the cups contained the bitter solution and four water, and he was asked to taste each and to separate them into bitter and tasteless groups.

If he was successful a weaker solution was tried in the same way and so on until a distinction became impossible; if, on the other hand, the attempt failed, the same procedure was repeated using a stronger solution. The *definite* threshold was then taken to be the lowest concentration at which perfect separation was achieved.

An additional rule was also observed, which was necessitated by the occasional occurrence of an almost perfect result, namely the correct separation of three pairs of cups and the confusion of one pair. In this case the athlete was offered one cup containing water and another the solution. If he could separate this pair correctly the solution number was accepted as his definite threshold but not otherwise.

[1] This value is of course not the same as the laboratory physiologists' threshold but is highly correlated with it. Its choice is a compromise between the practicalities of field tests and statistical requirements.

III. Results

According to the general procedure of this investigation the athletes were classified according to sex, race, and sport speciality. From this, four major groups of races (Caucasoid, Negroid, Mongoloid, and Mestizo) were constituted, as well as two groupings of sport specialities. The first of these groupings, a physiological one, distinguishes between *explosive sports* (weight lifting, throwing, boxing and jumping) *sports of speed* (running, up to 800 meters; swimming, up to 400 meters; cycling up to 4000 meters), *sports of skill* (gymnastics, pole vault, diving, pentathlon, modern pentathlon, decathlon, basketball, and wrestling), and *sports of endurance* (running 1500 meters and more; swimming 1500 meters for men and 800 meters for women; cycling long distances; water polo; rowing and canoeing); while the other grouping, a psychological classification, contains only two categories: *aggressive sports* (boxing, wrestling, basketball, water polo, cycling) and *nonaggressive sports* (all the others).

Table 5.1 shows the number of nontasters among the athletes who presented themselves for testing according to the four major races and the four physiological categories of sports. The data are listed separately for males and females.

TABLE 5.1. INCIDENCE OF NONTASTERS (LEFT UPPER CORNER); TOTAL OF PARTICIPANTS IN ANY PARTICULAR COMBINATION OF RACE AND SPORT SPECIALTY (LEFT LOWER CORNER); CALCULATED FREQUENCY OF NONTASTERS (RIGHT UPPER CORNER) AND ASTERISKS WHERE THE FREQUENCY SIGNIFICANTLY DEVIATES FROM THE FREQUENCY IN THE CONTROL POPULATION. * P = 0.05, ** P = 0.01, *** P = 0.001.

SPORT	CAUCASOIDS		NEGROIDS		MONGOLOIDS		MESTIZOS		TOTALS	
MALES										
EXPLOSIVE	20 120	0.167 *	6 65	0.092	1 23	0.043	2 47	0.043	29 255	0.113
SPEED	9 64	0.140 *	8 83	0.096	0 11	0.000	2 25	0.080	19 183	0.103
SKILL	19 149	0.128 ***	2 28	0.071	3 20	0.150	4 38	0.105	28 235	0.119
ENDURANCE	54 298	0.181 **	2 32	0.063	2 30	0.067	4 80	0.050 *	62 440	0.140
TOTAL	102 631	0.161 ***	18 208	0.086	6 84	0.071	12 190	0.063 **	138 1113	0.124
FEMALES										
EXPLOSIVE	4 14	0.283	0 7	0.000	0 0	-	0 4	0.000	4 25	0.160
SPEED	1 28	0.036 **	0 19	0.000	0 1	0.000	0 11	0.000	1 59	0.017
SKILL	4 25	0.160	1 5	0.200	0 0	-	0 9	0.000	5 39	0.128
ENDURANCE	2 19	0.105	0 0	-	0 2	-	0 3	0.000	2 24	0.083
TOTAL	11 86	0.127 **	1 31	0.032	0 3	0.000	0 27	0.000 *	12 147	0.082

To evaluate these results it is first necessary to discuss the nontaster frequencies in the general populations from which the athletes have been drawn. This can only be done with some confidence for the Caucasoids and the Mestizos and, accordingly, the significance of deviation from an assumed control population is only indicated for those two racial groups.

The nontaster frequency for the Caucasoids is taken to be 0.25, which is a fairly low value if one surveys the numerous papers dealing with various European and Caucasoid North American populations. For the Mestizos—mostly Mexicans—a nontaster frequency of 0.145 was chosen, which is in between a value of 0.143 previously found in a sample of 857 urban Mexicans (Kalmus *et al.,* 1964) and 0.148 found in the control population of 363 urban Mexicans, tested for this investigation and presented in Figure 5.1.

Among the Negroids who presented themselves for testing, very few came from the United States while the majority came from diverse parts of Africa, Latin America, and several Pacific and Atlantic Islands. The Asiatic group was similarly heterogenous and it would be hazardous to stipulate specific nontaster frequencies for the controls of either.

Inspection of Table 5.1 shows that in all classes where a control frequency is available there is a numerical deficiency of nontasters, which in the case of the totals of all sports is quite significant for Caucasoids and Mestizos of both sexes.

Altogether the nontaster frequency among the female athletes was significantly lower than in the male athletes, but this also applies to the Mexican control population and perhaps no far-reaching conclusions should be drawn from it. Nevertheless, considering the involvement of the thyroid with the PTC polymorphism mentioned earlier, and bearing in mind that women are much more prone to thyroid diseases than men, this point might be considered in future investigations.

The numerical deficiency of nontasters can be observed in 15 out of the 16 categories for which control values are available; but as the significance of these differences is largely a function of the sample sizes it would be hazardous to single out any particular group of sports for which to claim that nontasting is more disadvantageous than for the others.

The subdivision into aggressive and nonaggressive sports as defined earlier was not very revealing as far as the PTC polymorphism is concerned. The total sample was somewhat smaller because women did not compete in the "aggressive" sports. The overall numerical deficiency of nontasters among the Caucasoid and Mestizo athletes was of course in evidence as in the other grouping for sports, but the only sizeable groups were the "aggressive" and the "nonaggressive" Caucasoid males and the difference in nontasters frequency between the "aggressives" (37 nontasters out of 260) and "nonaggressives" (65 out of 371) was not significant, the chi^2 = 1.22,1 df.

IV. Discussion

The underrepresentation of nontasters among the athletes is capable of quite varied interpretations. One of them might be based on social stratification which in this case would mean that successful athletes are mainly recruited from population strata which due to social or historical reasons are somehow biologically isolated. While such an explanation probably applies to the differences in taster frequencies between races and castes and certainly applies to the difference between American Caucasoids and Negroids, it is difficult to see how it might work in the case of athletic performance operating in the same way in several different societies and populations.

Another explanation, which we are inclined to accept, presumes real physiological connections between the phenotype of an individual and his athletic capacities. Such connections are at present purely hypothetical. One of them, which as mentioned in the introduction helped to start this investigation, is that tasters and nontasters differ in some thyroid functions and are thus more or less capable to perform the various extreme tasks of the competitive Olympic athlete. Those differences may be ramified through the endocrine system or may directly impinge on the release of muscular or nervous energy. It must, however, be borne in mind that quite different developmental and physiological mechanisms may be interpolated between the presumably enzymatic difference of tasters and nontasters and their athletic performance.

Finally some evolutionary speculation may be permissible: Differences of taster frequencies between human populations, and, in fact, primate populations (Kalmus, 1969) have been connected with the rejection of food plants containing compounds of the same class of bitter substances to which PTC also belongs. The underrepresentation of nontasters among the athletes in our samples suggests another possible explanation, namely that homozygotes for the nontaster gene may be less fit to sustain some physical exertions needed by the prehistoric modes of life. The high frequency of nontasters in Caucasoid populations and other populations with a long history of agricultural and urban existence may be thus attributed to a relaxed selection against the nontaster gene as suggested for other genes by Post (1962). Nontasters would then in all populations be less capable of the extreme physical exertions common to the athletic and the nonurban life; and while the less urbanized people can draw their Olympic competitors from almost their entire population, the more urbanized cannot recruit many athletes from the considerable fraction of their population who are nontasters.

V. Summary

The number of nontasters for PTC in a sample of athletes at the XIXth Olympic Games at Mexico was altogether smaller than would be expected from their frequency in the populations from which they were drawn. For men and women of Caucasoid and Mestizo origin, for which control data exist, the significance of this numerical underrepresentation has been established and it seems to apply to most sport specialities.

Bibliography

Fox, A. L. (1932). The relationship between chemical constitution and taste. *Proc. U.S. Nat. Acad. Sci.* **18**, 115.

Harris, H., and Kalmus, H. (1949). The measurement of taste sensitivity to phenylthiourea. *Ann. Eugen. London* **15**, 24–31.

Harris, H., Kalmus, H., and Trotter, W. R. (1949). Taste sensitivity to phenylthiourea in goitre and diabetes. *Lancet,* 1949, 1038.

Kalmus, H. (1958). Improvements in the classification of the taster genotypes. *Ann. Hum. Genet.* **22**, 222–230.

Kalmus, H. (1962). Genetical taste polymorphism and thyroid disease. *Proc. 2nd Int. Congr. Hum. Genet. Rome* 1856–1862.

Kalmus, H. (1969). The senses of taste of chimpanzees and other primates. "The Chimpanzee." Vol. II. 190–199. S. Karger, Basel and New York.

Kalmus, H., and Maynard Smith, Sheila (1965). The antimode and lines of optimal separation in a genetically determined bimodal distribution, with particular reference to Phenylthiocarbamide sensitivity. *Ann. Hum. Genet.* **29**, 127–138.

Kalmus, H., de Garay, A. L., Rodarte, U., and Cobo, L. (1964). The frequency of PTC-tasting, hard ear wax, colour blindness and other genetical characters in urban and rural Mexican populations. *Hum. Biol.* **36**, 134–145.

Post, R. H. (1962). Population differences in red and green vision deficiency: A review and a query on selection relaxation. *Eugenics Quarterly* **9**, 31.

6

Single Gene Systems of Blood

I. Introduction

The athlete is a superbly functioning and complex being. It is not possible to approach the study of genetic influences on athletic ability through the classic use of variant genes which lead to easily definable defects. Rather, one is faced with the problem of comparing normal persons who excel in sports with those who do not or of comparing normal persons who excel in one sport with persons who excel in a different sport.

Performance of a person in any complex task is many steps removed from ordinary gene activity and is likely to reflect the coordinated—or uncoordinated—actions of many genes. However, better understanding of genetic influence on sports ability must move toward elucidating the role of limited numbers of genes.

The present studies have been based on two approaches. The anthropological studies are concerned with describing the physical attributes of athletes, in hopes of demonstrating that certain traits do or do not characterize sports ability. What the anthropologist measures is still far removed from primary gene action, but there is the compensation that many physical traits do appear to be largely under genetic control. Furthermore, the traits selected for study are likely to be relevant to athletic ability, even though the conclusions may not always be predicted.

The second approach involves the study of single genes. Studies of single genes are intellectually appealing in that the effects investigated can be more readily related to primary gene action. A major drawback is the inability to select for study those genes likely to be relevant to athletic ability. The main reasons for choosing one genetic locus for study rather than another are, first, the existence of variant alleles with favorable frequencies, and, second, the existence of convenient means of observing the genetic differences. The likelihood is rather great, therefore, that none of the systems studied will be associated with differences in athletic ability. We would accept as useful, however, the demonstration that a particular variation is not an important factor in sports ability. The frequency with which new gene differences are being reported in the literature assures the availability of many new systems for future studies.

The blood groups are favorable examples of metabolic products which can be treated as simple genetic traits. Understanding of the chemical basis of blood group specificity is limited to the ABO groups. In its simplest form—A versus non-A, and B versus non-B—the classification of persons according to ABO type presents only trivial problems. However, there are several subtypes of A and of B, apparently inherited, and these present much greater problems of classification and genetic analysis. Indeed, the subtypes behave in many respects as quantitative traits and are similar to other metabolic intermediates in responding to genetic and environmental influences.

The assessment of blood group genotypes is, therefore, one step further removed from the genes than are proteins. The number of alleles likely to be lumped together by limitations in technique is much greater than in the case of proteins, even though electrophoretic methods are capable of resolving only a portion of protein variants. By single-gene difference, therefore, is meant variation at a single gene locus, and the discussions of alleles always refer to distinctions possible in the laboratory, which may in some instances discriminate very crudely.

II. The Meaning of Genetic Variation

The number of genetic loci in man is not known but must be many thousands. The proportion of these which are variable has been a matter of concern, and some information is now available. First, it is important to distinguish between loci which are polymorphic and those which show rare nonpolymorphic variants. Polymorphism has been defined as the existence of two or more forms of a gene in frequencies which cannot be explained by mutation pressure alone (Ford, 1940). In practice, such a distinction will often be impossible, but it is generally accepted that an allele frequency of 1% or more is unlikely to occur purely by mutation pressure. The ability to mutate is an intrinsic characteristic of genes, and we may suppose that, no matter how stringent selection may be, some amino acid substitutions in a protein would function essentially as normal. Such mutations should remain rare unless there is a chance increase, through sampling or founder effects in very small proportions, or unless selection favors the mutant form.

Studies in man (Harris, 1966; Lewontin, 1967) and *Drosophila* (Hubby and Lewontin, 1966; Lewontin and Hubby, 1966) suggest that the frequency of randomly chosen loci which are polymorphic is of the order of 30%. Whether these loci are representative or not has not been settled (Gillespie and Kojima, 1968), but it seems likely that the proportion of loci which are polymorphic is large. A locus may be polymorphic in one small population but not in others. In this case, one would suspect the gene frequencies to be due largely to chance. On the other hand, there are loci such as those controlling the MNSs blood groups and the haptoglobins which are polymorphic in every human population. This suggests that selection favors the maintenance of the polymorphism.

The theoretical difficulty to the involvement of selection in so many loci has not been settled satisfactorily. This fact does not preclude studies of association of genetic traits with other attributes, and such studies may well provide important insight into how a few systems are maintained.

One consequence of the extensive genetic variability is that no two populations are identical in the kinds and frequencies of the different allelic forms of a gene. Certain alleles may be very limited in their distribution, occurring only in one or a few populations. Others are known in every population. The frequencies of the alleles of a particular genetic locus may vary, so that populations with the same array of alleles may nevertheless be distinguished by the relative frequencies of these alleles. These facts make it imperative that any study of gene associations take into account the population from which the subject originated.

III. Systems Selected for Study

Several considerations were involved in the choice of loci to be studied. From the genetic point of view, natural selection is more likely to be involved in maintaining a polymorphism if the locus is polymorphic in all or virtually all populations. Therefore, rare variants or variants limited to one group would be less likely to yield interesting results.

Another consideration was the number of persons to be studied. A rare variant, even if widespread geographically and even if it is associated with sports ability, is not likely to yield statistically significant results with a limited number of athletes. Therefore, it seemed better to give priority to those systems whose gene frequencies were more favorable.

From a physiological point of view, systems concerned with energy, metabolism and respiration would seem more promising than enzyme or protein systems not known to be directly involved in these areas of metabolism. The complexity of the metabolic network makes such selections no more than guesses, but several enzymes meeting other criteria were available for study.

One further consideration was the availability of suitable control data. Ideally, each athlete should have been matched with someone from the same ethnic background, etc., who lacks athletic potential. This clearly was not feasible, and the best that could be hoped for was to compare gene distributions among athletes to gene distributions from the parent population. This is not entirely satisfactory. However, if an apparent association is consistent in a number of groups, then the association is less likely to be due to ethnic stratification within the country.

Certain practical considerations also influenced the selection of systems for study. While many variants can be detected on gel electrophoresis (Smithies, 1955a) certain ones more nearly meet the requirement of the present study. From the technical side, it is important to make use of systems which are rapid and reliable. Further, some systems can be combined into a single procedure, greatly reducing the required time and effort. The systems ultimately selected were red cell glucose-6-phosphate dehydrogenase, acid phosphatase, and hemoglobin; and plasma haptoglobin, transferrin, and group-specific component.

The other technique which yields similar data on single gene systems is blood grouping. Certain of the blood group systems—ABO, MN, and Rh-Hr—meet the requirements as outlined above. The systems have high technical reliability and can be processed rapidly. While less can be inferred concerning gene structure, the extensive and effective use of these systems in population genetics is evidence of this utility.

IV. Statistical Analysis for the Study of Single Gene Systems

The great variability in gene frequencies among different populations makes necessary the comparison of frequencies among athletes with frequencies in a synthetic population designed to represent the population from which the athletes were drawn. The synthetic population was prepared for each genetic system, the proportion of English, Japanese, etc. being exactly the same as in the sample of athletes for whom information was available. Since there was an interest in analyzing the four main groups of sports specialities independently, four such comparison populations were actually constructed for each gene system. The different countries and racial groups were very differently represented among the sports specialties, so that the expected proportion of phenotypes would also differ among the sports specialties.

As described earlier, athletes from each country were classified into one of several major racial groups. In a number of instances there might be only one athlete for any country–race combination. The decision was made not to use data unless three or more persons had been studied for each country–race combination. This greatly reduced the computational effort, and the information lost was trivial.

Published gene frequencies for each population were used to establish the expected phenotype frequencies among the athletes. This was accomplished by the simple procedure of summing up the proportion of athletes from each population group multiplied by the phenotype frequencies published for that population group. A number of approximations were necessary, since published data are inadequate for many of the gene systems and populations. These approximations undoubtedly add to the departure from expectation observed, especially in those systems where gene frequencies are variable from one population to another. The general conformity of the observed phenotypes with the calculated expected phenotypes suggests that the approximations were reasonable for the most part. Nevertheless, the interpretation of the results must be tempered by the knowledge that the error of the approximations was potentially large in some instances.

V. Red Cell Antigens

Three red cell antigen systems were selected for study, the ABO, the MN, and the Rh-Hr. Selection was made on the basis of ease of typing and reliability of results.

The antisera and lectins used were all donated by the Hyland Laboratories of Los Angeles, California, and consisted of the following: antisera for A, B, AB, C, c, D, E, e, and human serum (for Coombs test). The lectins were for A_1, M, and N. The techniques used were those suggested in each instance by the manufacturer.

Samples of blood (6 milliliters) were drawn from each athlete by venipuncture using disposable syringes and needles. The blood was placed immediately in a 13 × 100 millimeter glass tube containing two drops of 10% EDTA as anticoagulant. It was stored at 4°C. All samples were further processed within 24 hours, the plasma being frozen for other studies. The typing of red cells for antigens was carried out within 24 hours after collection of the blood.

A. *ABO Groups*

The analysis for ABO groups was based on 657 athletes from 50 countries (Table 6.1). Although the A groups were subdivided into A_1 and A_2 during testing, they were pooled for the analysis. Control gene frequencies were taken primarily from Mourant *et al.* (1958), supplemented by unpublished results of A. E. Romero on Mestizos.

In selecting the gene frequencies to represent the parent population, a number of arbitrary decisions were made. In many cases there was a wide range of choices. Emphasis was given to those reports where the numbers were very large and where the subjects appeared to be representative of the country as a whole rather than a genetic isolate. Where more than one such report was available, values were selected which appeared to be compatible with the major reports and which were representative. In other cases the reports from a given country were not always based on adequate sample size and on a representative population. In those cases, arbitrary decisions were made based on the range of values reported and on published values from related populations in a different country. In a few instances, primarily in western

TABLE 6.1. COMPARISON OF PHENOTYPE FREQUENCIES OF THE ABO SYSTEM WITH CALCULATED EXPECTED FREQUENCIES

SPORTS SPECIALTY	PHENOTYPE					χ^2 (3 d.f.)
	A	B	AB	O	TOTAL	
			OBSERVED			
EXPLOSIVE	35	23	7	66	131	
SPEED	43	11	8	66	128	
SKILL	52	25	6	60	143	
ENDURANCE	89	28	3	135	255	
TOTAL	219	87	24	327	657	
			EXPECTED*			
EXPLOSIVE	44.65	19.54	5.54	60.87	130.60	3.52
SPEED	41.47	20.50	5.25	60.79	128.01	6.35
SKILL	52.78	19.67	6.13	64.34	142.92	1.75
ENDURANCE	99.33	29.59	9.82	116.26	255.00	8.92**
TOTAL	238.23	89.30	26.74	302.26	656.53	3.92

χ^2 (heterogeneity) = 16.62, 9 d.f., $P \geq .05$.
*Expected values were calculated primarily from Mourant et al. (1958); see text.
**$.05 > P > .02$.

Europe, where reported values from a number of countries differed by insignificant amounts, the populations were pooled to reduce the calculations.

Table 6.1 is a comparison of the expected and observed phenotype frequencies for the four classifications of sports specialty and for the total number of athletes. The nonsignificant chi square for the total fails to provide evidence that athletes differ in their phenotype frequencies from the populations from which they were drawn. The nonsignificant chi square for heterogeneity fails to provide evidence that the different sports specialties vary in this respect. The one significant chi square associated with sports involving endurance must, therefore, be attributed to chance variation. Inspection of the observed shows no consistent pattern of deviation. Altogether, this analysis provides no evidence to support a hypothesis that ABO blood group is important in developing athletic ability or an athletic specialty.

B. *MN Groups*

The results of 655 blood samples are given in Table 6.2 for the MN groups. It was hoped initially to include results on typing for S and s. However, some laboratory problems arose which were not recognized in time to permit the utilization of this system. The MN groups are, therefore, presented as a simple two-allele system.

TABLE 6.2. COMPARISON OF MN PHENOTYPE FREQUENCIES AMONG SPORTS SPECIALTIES WITH CALCULATED EXPECTED FREQUENCIES

SPORTS SPECIALTY	PHENOTYPE M	MN	N	TOTAL	χ^2 (2 d.f.)
		OBSERVED			
EXPLOSIVE	43	41	26	110	
SPEED	33	51	26	110	
SKILL	50	70	23	143	
ENDURANCE	105	127	60	292	
TOTAL	231	289	135	655	
		EXPECTED*			
EXPLOSIVE	35.85	51.95	22.18	109.98	4.38
SPEED	33.31	52.80	23.88	109.99	.19
SKILL	45.27	68.30	29.40	142.97	1.93
ENDURANCE	91.43	140.31	60.22	291.96	3.28
TOTAL	205.86	313.36	135.68	654.90	4.97

χ^2 (heterogeneity) = 4.81, 6 d.f., P > .05.
*Expected values were calculated from published gene frequencies for Caucasoid populations (Race and Sanger, 1962, p. 80; Mourant, 1954; and Constantoulis and Paidoussis, 1958) and for Filipinos (Fraser et al., 1964). Reported phenotype frequencies were used for American Negroids (Reed, 1968) and Mestizos (A. E. Romero, unpublished data).

The universal polymorphism for MN groups makes this an especially good candidate as a balanced polymorphic system to be maintained by selection. This would be reflected as an excess of heterozygous MN persons compared to homozygous M and N persons. Table 6.2 reveals the contrary trend. While none of the chi square values is significant, the observed numbers of MN persons in explosive and endurance type sports are substantially less than expected. Hence these data provide no support for outstanding physical ability as a factor in selection at the MN locus.

C. *Rh Groups*

In order to simplify the analysis of the Rh system, the three major factors have been considered separately. This discards some information, but any major deviation in Rh frequencies should be reflected in the individual factors. These results are presented in Tables 6.3, 6.4, and 6.5. As with the previous blood groups, there is no suggestion that general athletic ability or sports specialty is associated with Rh type.

VI. Red Cell Enzymes

The genetic study of red cell enzymes has been greatly stimulated in the past few years by the discovery of a number of genetic polymorphisms. The earliest known of these was glucose-6-phosphate dehydroge-

TABLE 6.3. COMPARISON OF Cc PHENOTYPE (Rh SYSTEM) FREQUENCIES AMONG SPORTS SPECIALTIES WITH CALCULATED EXPECTED FREQUENCIES

SPORTS SPECIALTY	PHENOTYPE C	Cc	c	TOTAL	χ^2 (2 d.f.)
		OBSERVED			
EXPLOSIVE	29	61	66	156	
SPEED	15	52	82	149	
SKILL	34	78	64	176	
ENDURANCE	55	166	98	319	
TOTAL	133	357	310	800	
		EXPECTED*			
EXPLOSIVE	26.31	64.87	64.81	155.99	.528
SPEED	20.93	57.66	70.41	149.00	4.144
SKILL	34.37	82.30	59.33	176.00	.424
ENDURANCE	65.45	152.84	100.71	319.00	2.875
TOTAL	147.06	357.67	295.26	799.99	2.081

χ^2 (heterogeneity) = 5.890, 6 d.f., P > .05.
*Expected values were calculated from gene frequencies taken from appropriate populations as published by Mourant (1954).

TABLE 6.4. COMPARISON OF D PHENOTYPE (Rh SYSTEM) FREQUENCIES AMONG SPORTS SPECIALTIES WITH CALCULATED EXPECTED FREQUENCIES

SPORTS SPECIALTY	PHENOTYPE D(+)	D(-)	TOTAL	χ^2 (1 d.f.)
		OBSERVED		
EXPLOSIVE	141	15	156	
SPEED	133	16	149	
SKILL	158	18	176	
ENDURANCE	280	39	319	
TOTAL	712	88	800	
		EXPECTED*		
EXPLOSIVE	137.46	18.53	155.99	.764
SPEED	132.17	16.83	149.00	.046
SKILL	152.51	23.48	175.99	1.477
ENDURANCE	272.78	46.21	318.99	1.316
TOTAL	694.92	105.05	799.97	.243

χ^2 (heterogeneity) = 3.360, 3 d.f., P > .05.
*Expected values were calculated from gene frequencies taken from appropriate populations as published by Mourant (1954).

TABLE 6.5. COMPARISON OF Ee PHENOTYPE (Rh SYSTEM) FREQUENCIES AMONG SPORTS SPECIALTIES WITH CALCULATED EXPECTED FREQUENCIES

SPORTS SPECIALTY	PHENOTYPE E	Ee	e	TOTAL	χ^2 (2 d.f.)
		OBSERVED			
EXPLOSIVE	5	48	103	156	
SPEED	6	33	110	149	
SKILL	8	44	124	176	
ENDURANCE	7	87	225	319	
TOTAL	26	212	562	800	
		EXPECTED*			
EXPLOSIVE	4.30	40.47	111.22	155.99	2.123
SPEED	3.86	38.16	107.00	149.02	1.968
SKILL	5.65	48.77	121.57	175.99	1.493
ENDURANCE	9.82	88.31	220.86	318.99	.907
TOTAL	23.63	215.71	560.65	799.99	3.512

χ^2 (heterogeneity) = 2.98, 6 d.f., P > .05.
*Expected values were calculated from gene frequencies taken from appropriate populations as published by Mourant (1954).

nase (reviewed in Giblett, 1969). This X-linked enzyme is polymorphic in many populations and has been related to resistance to malaria. The extensive data on glucose-6-phosphate dehydrogenase and in addition on acid phosphatase make these enzymes especially interesting for investigation in the present study.

A. *Glucose-6-Phosphate Dehydrogenase*

The investigation of glucose-6-phosphate dehydrogenase (G6PD) was limited to starch gel electrophoresis. This procedure recognizes electrophoretic variation only. Hence the two phenotypes, Gd A(+) and Gd A(−), commonly found in persons of African descent, are grouped together. An illustration of the Gd A and Gd B types is given in Figure 6.1. Table 6.6 gives the number of persons of each G6PD phenotype according to sports specialty compared to the calculated number of expected phenotypes. Analysis was the same as for previous systems. Only males are reported for this trait since there were too few females for analysis and females cannot be pooled with males for X-linked traits. No significant differences were found between athletes and the parent populations or among athletes in different sports specialties. This particular table points out the different racial composition of the sports specialties, Negroids being especially prevalent in sports involving speed, and emphasizes the need to make corrections in order to test even among the athletes themselves.

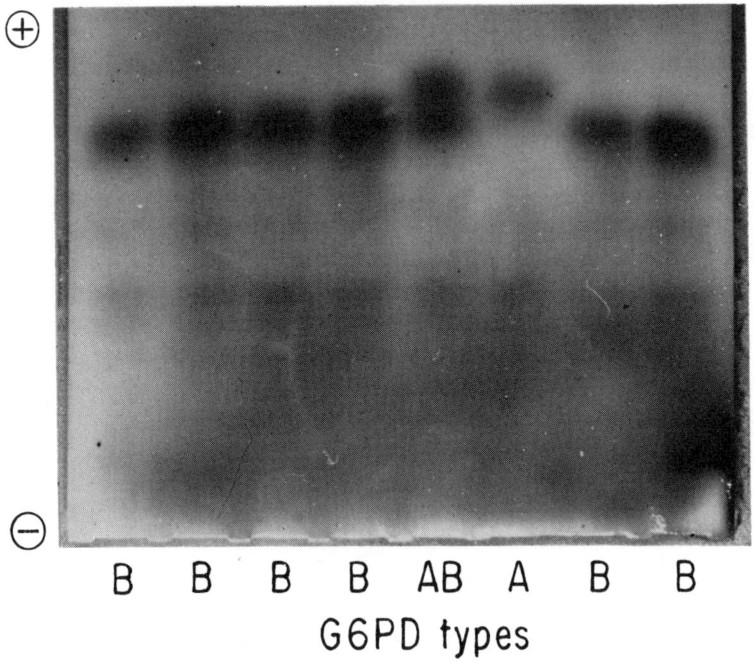

Figure 6.1 G6PD phenotypes as shown by starch gel electrophoresis. Samples were inserted at the bottom (cathodal end) and migrated toward the top (anode). Six of the samples are type B. Number 5 (from the left) is from a female and shows the AB heterozygous pattern; Number 6 is type A.

TABLE 6.6. COMPARISON OF THE PHENOTYPE FREQUENCIES OF THE GLUCOSE-6-PHOSPHATE DEHYDROGENASE SYSTEM WITH CALCULATED EXPECTED FREQUENCIES (MALES ONLY)

SPORTS SPECIALTY	PHENOTYPE A	B	TOTAL	χ^2 (1 d.f.)
		OBSERVED		
EXPLOSIVE	7	121	128	
SPEED	13	88	101	
SKILL	3	130	133	
ENDURANCE	6	280	286	
TOTAL	29	619	648	
		EXPECTED*		
EXPLOSIVE	11.29	116.71	128	1.788
SPEED	15.26	85.74	101	.394
SKILL	4.57	128.43	133	.559
ENDURANCE	7.38	278.62	286	.265
TOTAL	38.50	609.50	648	2.492

χ^2 (heterogeneity) = .514, 3 d.f., P > .05.
*Expected values were calculated on the basis of the following allele frequencies: Caucasoids, Mongoloids, and Mestizos, $\underline{Gd^A}$ = .00; African Negroids, $\underline{Gd^A}$ = .39 (Porter et al., 1964; Reys et al., 1970); American Negroids and others of mixed African ancestry, $\underline{Gd^A}$ = .29 (Yoshida et al., 1967).

B. *Acid Phosphatase*

Red cell acid phosphatase was classified by starch gel electrophoresis according to the method of Karp and Sutton (1967). With the system used, the products of the common allele P^a migrate toward the cathode, while the products of the other common allele, P^b, migrate toward the anode (Figure 6.2). The products of the allele P^c, which is common among Europeans, migrate identical to the B products, but with different band concentrations. A fourth allele, P^r, is found occasionally in persons of African descent. The R pattern, in combination with the B or C patterns, may be overlooked, and the samples may be classified as B or C.

Table 6.7 presents the analysis of acid phosphatase, distinguishing the common phenotypes but with the rare types pooled for statistical convenience. The analysis indicates a consistent difference from the expected values, with excess A and B homozygotes.

Because of the likelihood of incorrect classification of samples, these data were reanalyzed, treating acid phosphatase as a two-allele system, P^a and $P^{\text{non-}a}$. The A and non-A products migrate in opposite directions on electrophoresis, and it is very unlikely that samples were misclassified as (*1*) A, (*2*) A/non-A, or (*3*) non-A. The results of this analysis are in Table 6.8. Although the observations conform much more nearly to expectation, there still appear to be too few heterozygotes. The chi

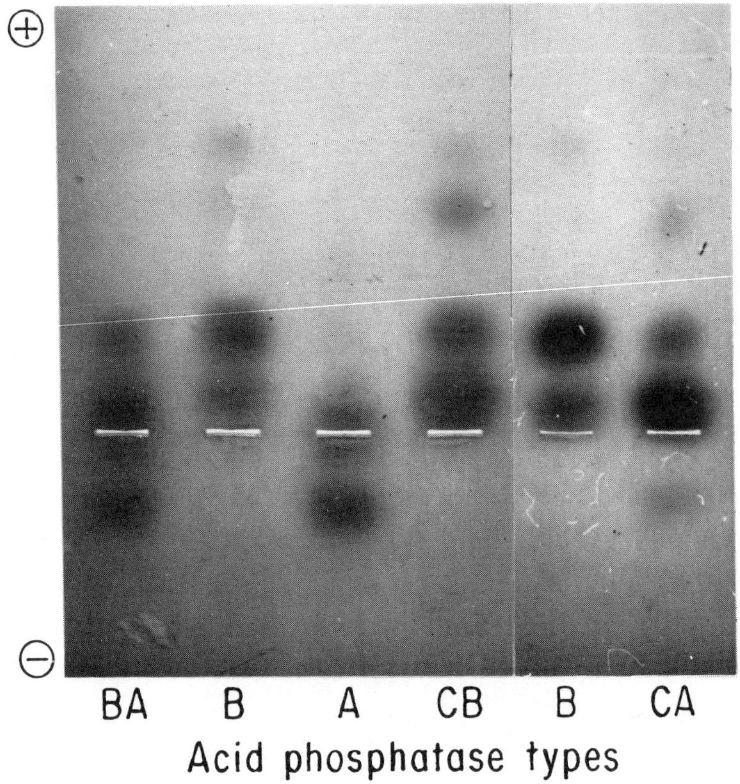

Figure 6.2 Common acid phosphatase phenotypes as shown by starch gel electrophoresis. The C pattern (which resembles CA but with no A bands) was not observed. In addition to the patterns illustrated, there were four samples classified as BR (which resembles B but with very heavy fast-moving satellite bands).

square values vary among sports specialties, but the test for heterogeneity gives a nonsignificant chi square, supporting the hypothesis that the difference is between athletes and control group.

At present, the most likely explanation of these findings is that the "control group" constructed from published studies is inappropriate. The data on distribution of acid phosphatase alleles are still very limited, and the values used here may not have been representative of all groups studied. The results in Table 6.9 support this explanation. An analysis similar to that reported in Table 6.8 was performed based only on Caucasoid athletes from Western European countries, Canada, New Zealand, and the United States, where allele frequencies are known with some confidence. None of the chi square values is significant. It is unlikely, therefore, that there is a biologically meaningful association between acid phosphatase phenotypes and sports ability in the other populations.

TABLE 6.7. COMPARISON OF RED CELL ACID PHOSPHATASE PHENOTYPE FREQUENCIES AMONG SPORTS SPECIALTIES WITH CALCULATED EXPECTED FREQUENCIES

SPORTS SPECIALTY	PHENOTYPE A	BA	B	CA	CB	BR	TOTAL	χ^2 (3 d.f.)
				OBSERVED				
EXPLOSIVE	16	55	81	1	0	1	154	
SPEED	15	44	89	0	1	1	150	
SKILL	13	51	83	0	2	1	150	
ENDURANCE	45	127	136	0	8	1	317	
TOTAL	89	277	389	1	11	4	771	
				EXPECTED*				
EXPLOSIVE	14.16	57.96	69.88		11.98#		153.98	10.47**
SPEED	12.61	55.00	71.42		10.94#		149.97	14.29***
SKILL	16.92	60.32	59.69		13.06#		149.99	19.20****
ENDURANCE	37.36	129.35	121.08		29.19#		316.98	17.41****
TOTAL	81.05	302.63	322.07		65.17#		770.92	53.96****

χ^2 (heterogeneity) = 7.41, 9 d.f., P > .05.
#CA + CB + BR; these types were combined in the analysis because of the small numbers of each type expected.
*Expected values were calculated from gene frequencies published in Giblett (1969, p. 436).
**P < .025. ***P < .005. ****P < .001.

TABLE 6.8. COMPARISON OF RED CELL ACID PHOSPHATASE PHENOTYPE FREQUENCIES AMONG SPORTS SPECIALTIES WITH CALCULATED EXPECTED FREQUENCIES. THE PHENOTYPES HAVE BEEN GROUPED TO FORM A TWO-ALLELE SYSTEM (P^a AND P^{non-a})

SPORTS SPECIALTY	PHENOTYPE A	BA + CA	B + BC + BR	TOTAL	χ^2 (2 d.f.)
			OBSERVED		
EXPLOSIVE	16	56	82	154	
SPEED	15	44	91	150	
SKILL	13	51	86	150	
ENDURANCE	45	127	145	317	
TOTAL	89	278	404	771	
			EXPECTED*		
EXPLOSIVE	14.16	61.55	78.33	154	.91
SPEED	12.61	57.88	79.51	150	5.44
SKILL	16.92	64.86	68.22	150	8.50**
ENDURANCE	37.36	119.93	139.71	317	2.96
TOTAL	81.05	304.22	365.77	771	11.36***

χ^2 (heterogeneity) = 6.45, 6 d.f., P > .05.
*Expected values were calculated from gene frequencies published in Giblett (1969, p. 436).
** P < .025. ***P < .005.

TABLE 6.9. COMPARISON OF RED CELL ACID PHOSPHATASE PHENOTYPE FREQUENCIES AMONG SPORTS SPECIALTIES, BASED ON A TWO ALLELE SYSTEM AND ON SELECTED CAUCASOID ATHLETES ONLY

SPORTS SPECIALTY	PHENOTYPE A	BA + CA	B + C + BC	TOTAL	χ^2 (2 d.f.)
			OBSERVED		
EXPLOSIVE	10	16	16	42	
SPEED	5	20	11	36	
SKILL	7	16	21	44	
ENDURANCE	23	68	56	147	
TOTAL	45	120	104	269	
			EXPECTED*		
EXPLOSIVE	5.75	19.58	16.67	42.00	3.82
SPEED	4.93	16.79	14.29	36.01	1.37
SKILL	6.02	30.51	17.47	44.00	1.86
ENDURANCE	20.12	68.53	58.35	147.00	.51
TOTAL	36.82	135.41	106.78	269.01	2.12

χ^2 (heterogeneity) = 5.44, 6 d.f., P > .05.

*Only Caucasoid athletes from western European countries, Canada, New Zealand, and the United States were included. Gene frequencies used were $\underline{p^a}$ = .37, $\underline{p^b}$ = .59, and $\underline{p^c}$ = .04 (based on studies summarized in Giblett, 1969, p. 436).

VII. Red Cell and Plasma Proteins

Gel electrophoresis can be used to detect a number of proteins directly with protein stains. Those in the plasma include haptoglobins (Hp), transferrins (Tf), and group specific component (Gc). Both Hp and Gc groups are highly polymorphic in every population investigated, while transferrin is polymorphic in several very large populations. These three proteins have been investigated extensively in many populations, making them good candidates for the present study. Hemoglobin also exists in many molecular forms and is polymorphic in several populations. As with the three plasma proteins listed earlier, gel electrophoresis is a convenient means of recognizing variants.

A. *Haptoglobin*

The haptoglobin polymorphism was first recognized by Smithies in 1955 (Smithies, 1955b; Smithies and Walker, 1955). (See Sutton, 1970, for a review.) Two common alleles were noted to occur in every population, although the frequency of the two alleles varies. In Negroids but rarely in other populations, persons are found who lack recognizable haptoglobin. These are given the phenotype designation Hp 0. Ordinarily such persons are ignored in calculating frequencies of the Hp^1 and Hp^2 alleles. This is, therefore, a small source of error potentially, but one which seems not to have been important in the present study. Similarly, a modified heterozygote phenotype, Hp 2-1 M, is found among Africans and persons of African descent. These persons can be pooled with the more usual Hp 2-1 heterozygotes for the purposes of analysis (Giblett and Steinberg, 1960; Sutton and Karp, 1964).

The method used for classifying haptoglobin types was based on acrylamide gel electrophoresis. A photograph of a typical gel is given in Figure 6.3. In Table 6.10 the observed haptoglobin frequences are compared with the calculated expected frequences. The agreement is very good, reflecting in part the excellent information available on the populations from which the athletes were drawn. These data, therefore, provide no suggestion that haptoglobin type is associated with athletic ability or sport specialty.

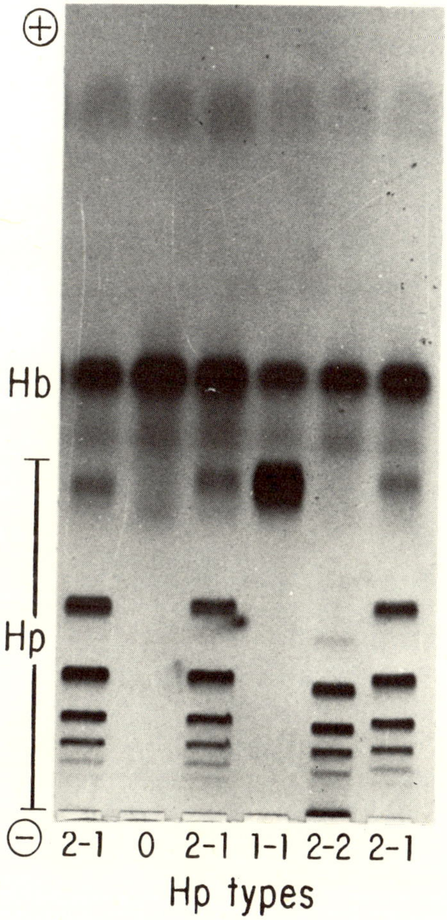

Figure 6.3 Common haptoglobin (Hp) phenotypes, using acrylamide gel electrophoresis and a peroxidase stain (*o*-dianisidine). The nonvariant intensely staining band is hemoglobin (Hb).

TABLE 6.10. COMPARISON OF Hp PHENOTYPE FREQUENCIES AMONG SPORTS SPECIALTIES WITH CALCULATED EXPECTED FREQUENCIES

SPORTS SPECIALTY	PHENOTYPE 1-1	2-1	2-2	0	TOTAL	χ^2 (2 d.f.)
			OBSERVED			
EXPLOSIVE	29	76	46	4	155	
SPEED	35	68	44	4	151	
SKILL	29	87	60	1	177	
ENDURANCE	61	148	112	4	325	
TOTAL	154	379	262	13	808	
			EXPECTED*			
EXPLOSIVE	34.80	73.75	46.44	-	155	1.039
SPEED	38.35	71.91	40.75	-	151	.764
SKILL	33.75	84.88	58.55	-	177	.711
ENDURANCE	59.12	155.39	110.48	-	325	.432
TOTAL	166.02	385.93	256.22	-	808	1.100

χ^2 (heterogeneity) = 1.85, 6 d.f., $P > .05$.
*Gene frequencies are summarized by Kirk (1968a) and Giblett (1969).

B. *Transferrin*

The transferrin which is common to all human populations and which is the predominant form in all populations is designated transferrin C (Tf C). Of the some 20 electrophoretic variants of transferrin which are known, only two (Tf D_1 and Tf D Chi) attain polymorphic frequencies. The Tf D_1 occurs with frequencies as high as 10% in many African populations and with frequencies of several percent in American Negroids and other descendant populations. Tf D_1 occurs at even higher frequencies among certain Australian aborigines. The other high-frequency variant, Tf D Chi, occurs widely among Oriental populations at frequencies of 4–5%. The geographic distribution of transferrin variants has recently been tabulated by Kirk (1968b).

The function of transferrin is to transport iron from the intestine, where it is absorbed, to various tissues, such as liver, where it is stored, and to the tissues where it is required for synthesis of iron-containing proteins. Transferrin is multifunctional in the sense that it not only has specific iron-binding sites but also can bind to reticulocytes in order to release its iron. It is, therefore, a very important protein in maintaining adequate synthesis of respiratory enzymes, many of which depend on iron for their crucial function. Some 20 variants of transferrin have been identified in human beings, and many variants are known among other mammals. So far as is known, these variants all function equally well in the transport of iron, although few have been tested completely for functions other than formation of the iron complex. The chemistry, physiology, and genetics of transferrin have been reviewed by Giblett (1969) and by Sutton and Jamieson (1970).

Transferrin can be readily seen on gel electrophoresis, since it is the major part of the beta globulin fraction. In the present study, classification of transferrin types was made on acrylamide gels. The observed transferrin phenotypes among the athletes are given in Table 6.11. The distribution of phenotypes is approximately as expected on the basis of known frequencies of variants in different populations.

TABLE 6.11. DISTRIBUTION OF TRANSFERRIN PHENOTYPES AMONG SPORTS SPECIALTIES

SPORTS SPECIALTY	PHENOTYPE			
	C	CD	CB	TOTAL
EXPLOSIVE	148	5	1	154
SPEED	143	5	1	149
SKILL	172	1	4	177
ENDURANCE	314	4	5	323
TOTAL	777	15	11	803

C. *Group Specific Component*

The Gc groups were first recognized by immunoelectrophoresis (Grubb and Laurell, 1956). Of special interest is the fact that the Gc groups are polymorphic in all populations examined, suggesting the action of selection in maintaining the two common alleles, Gc^1 and Gc^2. In this study, Gc types were determined by acrylamide gel electrophoresis. Examples of the common phenotypes are shown in Figure 6.4.

A comparison between the observed phenotype frequencies among athletes and the expected phenotype frequencies based on their parent populations shows no significant statistical difference, and fails to support the proposal that the Gc group is important to the development of superior athletic ability.

D. *Hemoglobin*

Of all the inherited protein variations, those involving hemoglobin are by far the best known and widely studied. Some 80 hemoglobin variants have been described, and many of them have been characterized biochemically. The common form of hemoglobin is designated Hb A.

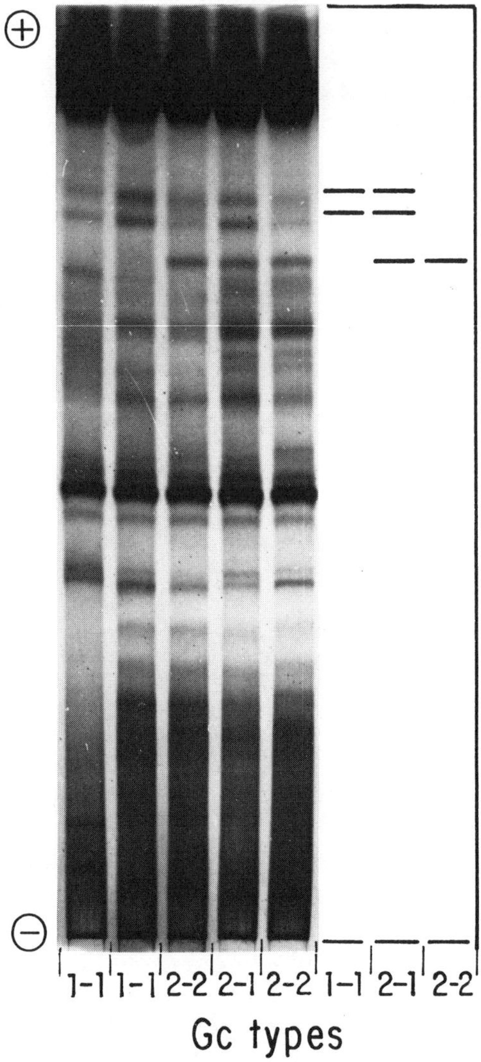

Figure 6.4 Group specific component types (Gc) using acrylamide gel electrophoresis. By using a general protein stain, many other proteins also are visible.

Only three of the variants, Hb S, Hb C, and Hb E, are found in polymorphic frequencies. The Hb S is found generally among Negroids, although the frequency depends strongly on the part of Africa from which the population originated. The same is true of Hb C, although the distribution in Africa is much more circumscribed. The Hb E is found in high frequency in southeast Asia (reviewed in Giblett, 1969).

The major variant hemoglobins appear to function normally in the transport of oxygen. They are not synthesized as rapidly as hemoglobin A, however, leading to some degree of anemia. In addition, hemoglobin S is much less soluble in the deoxygenated form, giving rise to crystalline-like structures inside the red cell which cause rupture of the cell and release of hemoglobin and other cell contents into the plasma. In the homozygous form, Hb S is responsible for sickle-cell disease, a generally lethal condition.

In view of the importance of hemoglobin in the transport of oxygen to tissues, one might expect very slight differences in function or amount of hemoglobin to be reflected in athletic potential. Especially in the case of hemoglobin S, one might suppose that the great oxygen demand which accompanies certain athletic activities might cause a certain amount of *in vivo* sickling of red cells even though this is not observed in heterozygotes under other conditions. Such persons might, therefore, be at a disadvantage, and even a small disadvantage would be expected to prevent such persons attaining Olympic status. It was surprising to discover that this was not the case and, indeed, a sizable number of Negroid Olympic athletes manifested the sickle-cell trait (Table 6.12, phenotype AS).

TABLE 6.12. DISTRIBUTION OF HEMOGLOBIN PHENOTYPES AMONG SPORTS SPECIALTIES

SPORTS SPECIALTY	PHENOTYPE			
	A	AS	AC	TOTAL
EXPLOSIVE	145	2	3	150
SPEED	134	8	2	144
SKILL	158	0	1	159
ENDURANCE	304	1	2	307
TOTAL	741	11	8	760

Hemoglobins were classified on the glucose-6-phosphate dehydrogenase gels, being clearly visible after electrophoresis without additional staining (Figure 6.5).

Table 6.12 shows the observed frequencies of hemoglobin types among the sports specialties. Because of the small number of variants, no formal analysis has been made. The observed numbers correspond, however, to expectation based on known frequencies of variants and composition of the populations comprising the sports specialties.

One special problem is raised by the high altitude in which the Olympic games were carried out in 1968. One could imagine that a greater oxygen deficit would be associated with this altitude and that persons

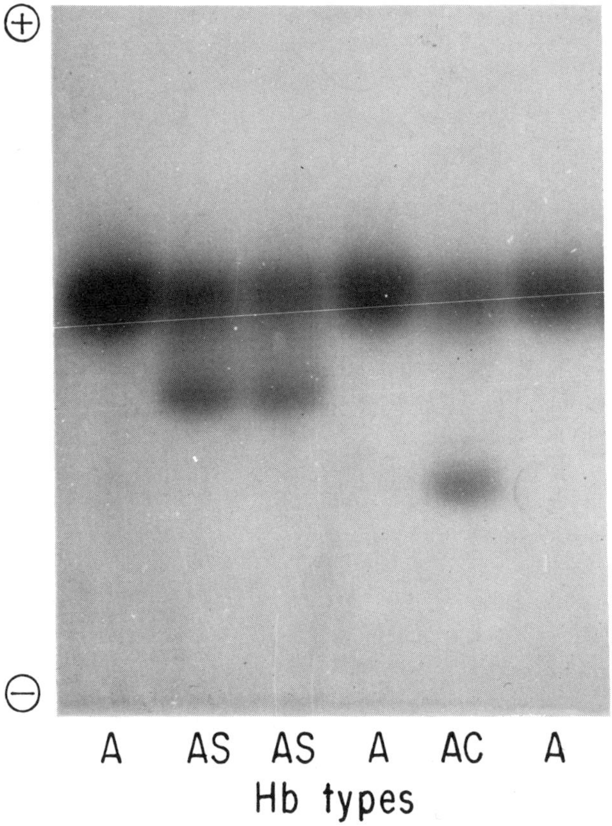

Figure 6.5 Hemogoblin types as shown on starch gel.

heterozygous for hemoglobin S would be more likely to form sickle cells in vivo than at lower altitudes. A crude test of such hemolysis can be made by noting the haptoglobin levels of the athletes following competition. Since sickling of red cells causes hemolysis and release of free hemoglobin into the plasma, the circulating haptoglobin is tied up by the free hemoglobin and removed from circulation. Therefore, a very low level of haptoglobin is one consequence of intravascular hemolysis, although this is not the only cause of low levels of haptoglobin. Blood was collected from one of the hemoglobin S heterozygotes shortly after competition. Although the level of haptoglobin was not assayed by standard quantitative methods, inspection of the gel and comparison to other samples collected at the same time failed to suggest a decreased level of haptoglobin. It is clear, therefore, that no extensive hemolysis occurred as a result of great physical stress at this high altitude.

VIII. Discussion

The failure to note an association of athletic ability or sports specialty with the single gene systems tested was largely as expected. The likelihood of selecting a single gene system which is associated with athletic ability is rather small. However, there are several considerations which made such a search worthwhile and which suggest continued effort in this direction.

That athletic ability is indeed associated with specific genetic markers is an assumption which needs little proof. However important training and motivation may be to the athlete, and certainly these are very important, there must exist a biological potential to be an athlete in order to achieve success. Within the normal population, many persons lack the biological potential to become an outstanding athlete, even though training and motivation may compensate in part for the biological limitation.

Whatever biological limitations may exist, they are almost certainly largely genetic in origin. These limitations must be attributed ultimately therefore, to the action of a series of single genes. One may reasonably expect to find associations between athletic ability and single genes if the search is continued long enough.

The systems selected for study were dictated largely by the necessity to work with systems which require minimal effort and which could be processed rapidly, in view of the large number of athletes to be investigated. Further, at the time the present study was planned, the number of polymorphic protein and enzymes known was limited. This situation is changing rapidly, with new polymorphic enzyme systems being reported frequently. Of particular interest are those systems related to energy metabolism, where one might expect small differences in normal function to become important under the severe stress of athletic competition. Enzymes such as 6-phosphogluconate dehydrogenase, phosphoglucomutase, phosphohexose isomerase, and glutathione reductase are examples of polymorphic systems which will be interesting for future investigations.

IX. Summary

Tests for association between athletic ability (including sports specialty) and various single gene systems were carried out. The genetic systems investigated were the ABO, MN, and Rh red cell types; haptoglobin, transferrin, and group specific component from plasma; and hemoglobin, glucose-6-phosphate dehydrogenase, and acid phosphatase variants of red cells. The data provide no support for the suggestion that any of these systems is associated with athletic ability or with sports specialty. However, an unexpected and exciting discovery was the presence of sickle-cell gene carriers among the Olympic athletes.

Bibliography

Constantoulis, N. C., and Paidoussis, M. (1958). The distribution of ABO, MNS, and Rh blood groups in Greece. *Vox Sang.* **3**, 145–154.

Ford, E. B. (1940). Polymorphism and taxonomy. *In* "The New Systematics" (J. Huxley, ed.). Oxford Univ. Press (Clarendon), London and New York.

Fraser, G. R., Giblett, E. R., Stransky, E., and Motulsky, A. G. (1964). Blood groups in the Philippines. *J. Med. Genet.* **1**, 107–109.

Giblett, E. R. (1969). "Genetic Markers in Human Blood" Blackwell, Oxford.

Giblett, E. R., and Steinberg, A. G. (1960). The inheritance of serum haptoglobin types in American Negroes: Evidence for a third allele, Hp^{2m}. *Am. J. Hum. Genet.* **12**, 160–169.

Gillespie, J. H., and Kojima, K-i. (1968). The degree of polymorphisms in enzymes involved in energy production compared to that in nonspecific enzymes in two *Drosophila ananassae* populations. *Proc. Natl. Acad. Sci.* **61**, 582–585.

Grubb, R., and Laurell, A. B. (1956). Hereditary serological human serum groups. *Acta Path. Microbiol. Scand.* **39**, 390.

Harris, H. (1966). Enzyme polymorphisms in man. *Proc. Roy. Soc. Lond.* [B], **164**, 298–310.

Hopkinson, D. A., Spencer, N., and Harris, H. (1964). Genetical studies on human red cell acid phosphatase. *Am. J. Hum. Genet.* **16**, 141–154.

Hubby, J. L., and Lewontin, R. C. (1966). A molecular approach to the study of genic heterozygosity in natural populations. I. The number of alleles at different loci in *Drosophilia pseudoobscura. Genetics* **54**, 577–594.

Karp, G. W., Jr., and Sutton, H. E. (1967). Some new phenotypes of human red cell acid phosphatase. *Am. J. Human Genet.* **19**, 54–62.

Kirk, R. L. (1968a). "The Haptoglobin Groups in Man." Monographs in Human Genetics, Vol. 4 (L. Beckman and M. Hauge, eds.). Karger, Basel.

Kirk, R. L. (1968b). World distribution of transferrin variants and some unsolved problems. *Acta Genet. Med. Gemel.* **17**, 613–640.

Kirkman, H. N., and Hendrickson, E. M. (1963). Sex-linked electrophoretic difference in glucose-6-phosphate dehydrogenase. *Am. J. Hum. Genet.* **15**, 241–258.

Lewontin, R. C. (1967). An estimate of average heterozygosity in man. *Am. J. Hum. Genet.* **19**, 681–685.

Lewontin, R. C., and Hubby, J. L. (1966). A molecular approach to the study of genic heterozygosity in natural populations. II. Amount of variation and degree of heterozygosity in natural populations of *Drosophila pseudoobscura. Genetics* **54**, 595–609.

Mourant, A. E. (1954). "The Distribution of Human Blood Groups." Blackwell, Oxford.

Mourant, A. E., Kopec, A. C., and Domaniewska-Sobczak, K. (1958). "The ABO Blood Groups: Comprehensive Tables and Maps of World Distribution." Blackwell, Oxford.

Porter, I. H., Boyer, S. H., Watson-Williams, E. J., Adam, A., Szeinberg, A., and Siniscalco, M. (1964). Variation of glucose-6-phosphate dehydrogenase in different populations. *Lancet* **1**, 895–899.

Race, R. R., and Sanger, R. (1962). "Blood Groups in Man" (4th ed.). Blackwell, Oxford.

Reed, T. E. (1968). Distributions and tests of independence of seven blood group systems in a large multiracial sample from California. *Am. J. Hum. Genet.* **20**, 142–150.

Reys, L., Manso, C., and Stamatoyannopoulos, G. (1970). Genetic studies on Southeastern Bantu of Mozambique. I. Variants of glucose-6-phosphate dehydrogenase. *Am. J. Hum. Genet.* **22**, 203–215.

Smithies, O. (1955a). Zone electrophoresis in starch gels: Group variations in the serum protein of normal human adults. *Biochem. J.* **61**, 629–641.

Smithies, O. (1955b). Grouped variations in the occurrence of new protein components in normal human serum. *Nature* **175**, 307–308.

Smithies, O. (1959). An improved procedure for starch-gel electrophoresis: Further variations in the serum proteins of normal individuals. *Biochem. J.* **71**, 585–587.

Smithies, O., and Walker, N. F. (1955). Genetic control of some serum proteins in normal humans. *Nature* **176**, 1265–1266.

Sutton, H. E. (1970). The haptoglobins. *Prog. Med. Genet.* **7**, 163–216.

Sutton, H. E., and Jamieson, G. A. (1970). Plasma glycoproteins: transferrin, haptoglobin, and ceruloplasmin. *In* "Glycoproteins" (A. Gottschalk, ed.). Elsevier, Amsterdam. (in press).

Sutton, H. E., and Karp, G. W., Jr. (1964). Variations in heterozygous expression at the haptoglobin locus. *Am. J. Hum. Genet.* **16**, 419–434.

Yoshida, A., Stamatoyannopoulos. G., and Motulsky, A. (1967). Negro variant of glucose-6-phosphate dehydrogenase deficiency (A-) in man. *Science* **155**, 97–99.

7

Dermatoglyphic Studies: Finger and Palm Prints

I. Introduction

In recent years the study of finger and palm prints has become an important method of research and gradually a large number of works have accumulated.

Examples of personal identification by finger-print patterns can be found in early Chinese documents. In the Western world, Grew (1684), Bidloo (1685), and Malpighi (1686), observed ridge patterns, but their investigations were restricted to direct observations on finger prints, simply noting similarities or differences in the ridge formation of individuals. Even as late as the 18th century, very little systematic work was done. It was not until 1823 that Purkinje attempted the first classification of finger prints. A system of using finger prints for personal identification was published by Faulds in 1880.

Our knowledge of the patterns formed by the papillary ridges on human finger prints has its foundation in the research of Francis Galton who pioneered in fundamental finger-print studies concerned with morphology, classification, inheritance, and racial variation. His findings were reported in a series of papers which are reviewed in his book, *Finger Prints* (1892). A great amount of literature from different countries preceding and following Galton's publications has been published.

II. Characteristics of Dermatoglyphics

The ridges of the epidermis and their arrangements are studied by taking prints of the areas concerned: finger and palm prints, and sole and toe prints.

The dermal ridge formation takes place during the third and fourth prenatal months. When complete, the ridges and the patterns formed by them remain unchanged during the individual's life. It is during this period of differentiation that factors affecting the growth of the limbs, including genetical ones, are likely to disturb the pattern formation.

In the study of dermal ridges, the "deltas" or "triradii" are important. A triradius can be defined as the meeting point of three spokes that demarcate three regions, each of which contains a system of almost parallel ridges. Triradii are found on fingers, palms, soles, and toes.

A. *Finger Prints*

1. *Pattern Type*

For convenience the digital patterns are here classified into three main types: arch, loop, and whorl. This classification (Galton, 1892) is based on the number of triradii present on the finger tip. The absence of a triradius gives rise to a simple arch, a loop has one triradius, and the whorl includes double loops and all patterns with two, sometimes three triradii.

2. *Pattern Index Intensity*

The distribution of the pattern form on the 10 fingers in the different populations tested has shown different values for each finger. In order to evaluate the pattern frequencies, the pattern intensity index is used (Cummins and Steggerda, 1935). It is calculated either as the average number of triradii per finger or the average number of triradii per individual. This index can be estimated for any group from the frequencies of whorls and loops, for each whorl has two triradii and each loop one. The values obtained per individual range from no triradius (all arches) to 20 triradii (all whorls).

3. *Ridge Count*

The quantitative approach to dermatoglyphics has led to another method of studying finger prints. On each print the ridges between two specified points are counted. Galton in 1895 originated the ridge-count technique, restricting it to loop subclassification. Henry (1901) used this system for identification purposes. The method consists of counting the number of papillary ridges in each pattern which cut or touch a straight line between the triradius and the core or center. Bonnevie (1924) extended the method to all types of digital patterns. The ridges are not formed in a completely regular way, some of them are interrupted, others branched. For simple arches (no triradius) the count is zero; in a loop there is one value (one triradius), while for the whorls (two triradii), there are two lines of count, one from each triradius to the center (or centers) of the whorl. At present only the higher count is used in the case of whorls (Figure 7.1). The ridge count gives a measure of pattern size which is independent of age.

This quantitative method provides a good basis for statistical analysis. The sum of the values of the ridge counts on the 10 fingers, that is

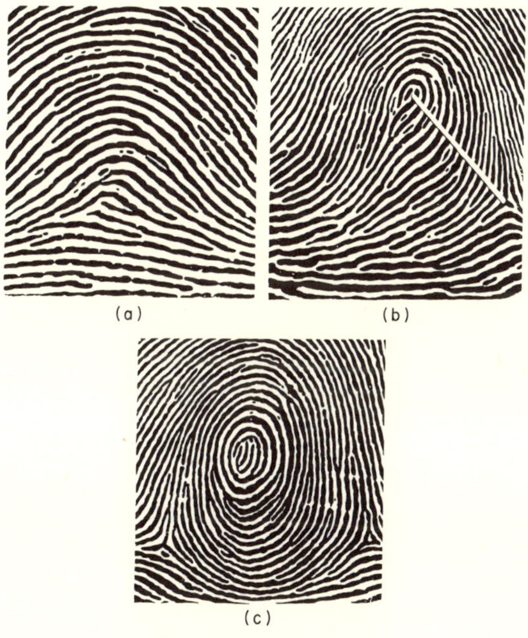

Figure 7.1 The three main patterns: (a) simple arch (no triradius) 0 ridge count; (b) loop (one triradius), one ridge count; and (c) whorl (two triradii) two ridge counts, only the highest is taken (Holt, 1961b).

the *total ridge count,* gives a single value for each individual. The total ridge count provides a new approach to the problem of the genetics of finger-print patterns. It has been extended to clinical and medical use, particularly in cases of chromosomal aberrations, as it is a very stable characteristic not affected by age.

The study of dermatoglyphics may also give important data in anthropological studies. Data from numerous population samples have shown racial differences. One of the most important anthropological differences is that between the sexes. In females the arch/whorls index is nearly always higher than in males (Dankmeijer, 1938), because females usually have more arches than males on fingers. There are also bimanual differences, the hands of one person never being exactly alike. On fingers pattern type (arch, loop, whorl) and pattern size (ridge count) frequencies have been studied in numerous ethnic groups. All show differences in traits between each other and between the sexes. The normal variations of pattern frequencies between different populations clearly indicate that such variations represent mainly hereditary differences.

B. *Palm Prints*

Palmar dermatoglyphics, like finger prints, differ between individuals of the same racial stock in the frequences of particular traits in ridge breadth and in detailed structure of the ridges (minutiae).

In the study of palm prints, a human palm is generally divided into five areas: hypothenar, thenar/I interdigital, and II, III, and IV interdigitals. These are the areas where patterns may occur (Figure 7.2) and various configurations may be observed in any of them.

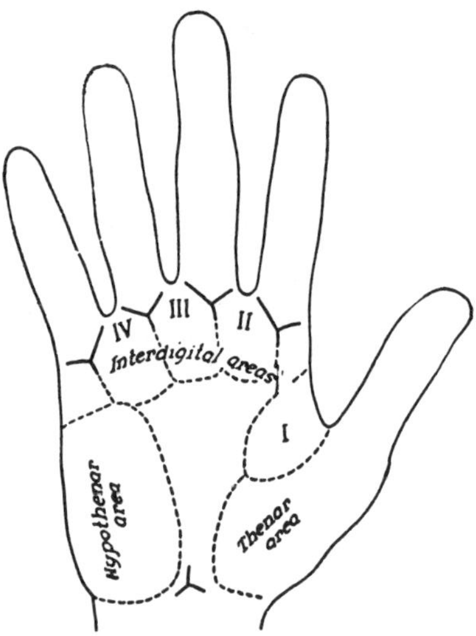

Figure 7.2 The five palmar configurations (Cummins and Midlo, 1961).

These configurations range in form from whorls, loops, arches, and vestiges to the total absence of patterns, called open fields. Whorls are similar to those on fingers, but often larger, sometimes with more than two triradii; vestiges represent irregularities in arrangement of the ridges, but are not associated with triradii. (Figure 7.3).

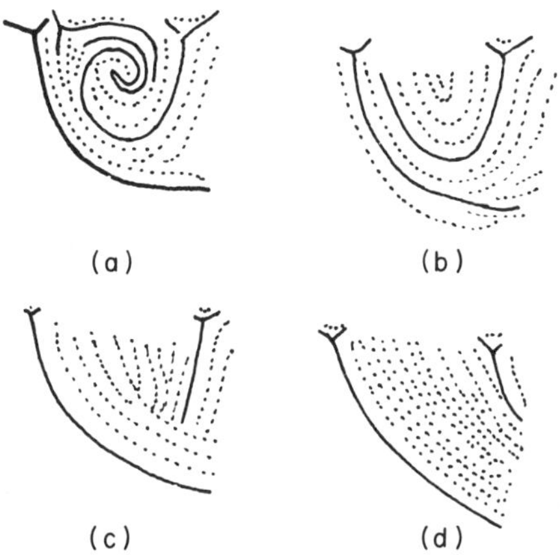

Figure 7.3 Different configurations on a palm: (a) whorl, (b) loop, (c) vestige, and (d) open field.

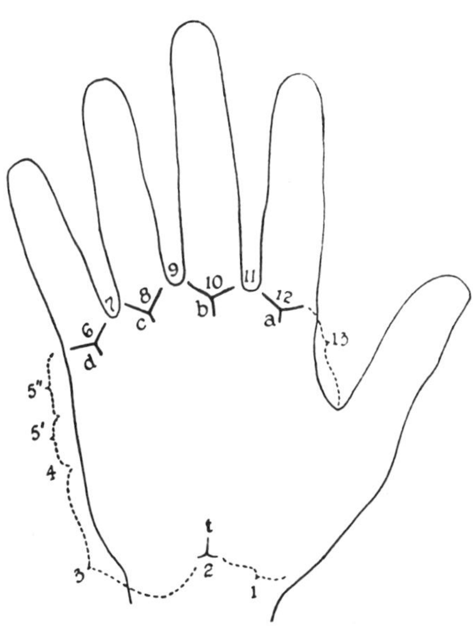

Figure 7.4 Digital triradii of the palm a, b, c, and d, and the marginal numbers of the palm to classify main lines (Cummins and Midlo, 1961).

1. *Digital Triradii*

In order to divide the interdigital areas, one takes into account the four triradii in the distal palm that are found normally below each finger from the index to the little finger. These are called digital triradii and are given the letters a, b, c, and d (Figure 7.4).

2. *Axial Triradius*

The axial triradius, called *t*, is generally found in the proximal region of the palm in normal populations. Its position is variable, but in the great majority of people it occurs at the base of the palm, in the depression between the thenar and hypothenar areas. There may be more than one axial triradius. When the triradius lies near the center of the palm it is termed *t″* and when it occupies an intermediate position between *t* and *t″* it is called *t′* (Figure 7.5).

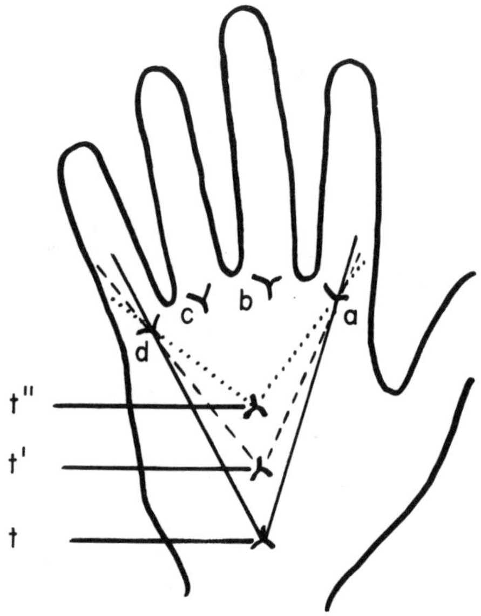

Figure 7.5 The axial *atd* angle and its three main positions.

3. *Main Lines*

From the proximal radiant of each of the digital triradii a, b, c, and d, a line is drawn on a palm print following the course of the ridges across the palm. These are the "main lines" and are designated by the capital letters A, B, C, and D, according to the triradius of origin. The main line T is drawn from the axial triradius *t* (Figure 7.6). Sometimes

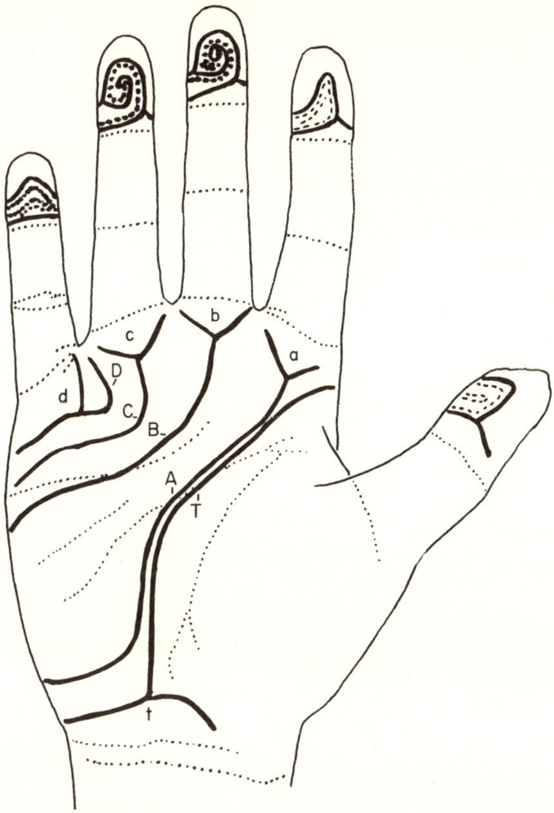

Figure 7.6 Main lines A, B, C, D, and T from the digital triradii a, b, c, d, and t.

one of the digital triradii is duplicated. Occasionally a digital triradius is absent and a main line cannot be drawn. This abortive state may be due to racial differences.

The margin of the palm is represented by the numbers that range from 1 to 13 and the point of exit of each main line from the palm can be described by a number. This quantitative method is useful for comparing populations (see Figure 7.4).

4. *Main-Line Index*

The terminal position of the main lines (A, B, C, and D) indicate the ridge direction of the palm. Lines B and C are relatively unimportant because they are limited by the extreme lines A and D. The terminations of lines A and D combined gives rise to the "main-line index" (Cummins, 1936). To determine this, another set of numbers around the periphery of the palm is used. Its value will reflect the general course of ridges either in a transverse or a longitudinal directions (Figure 7.7).

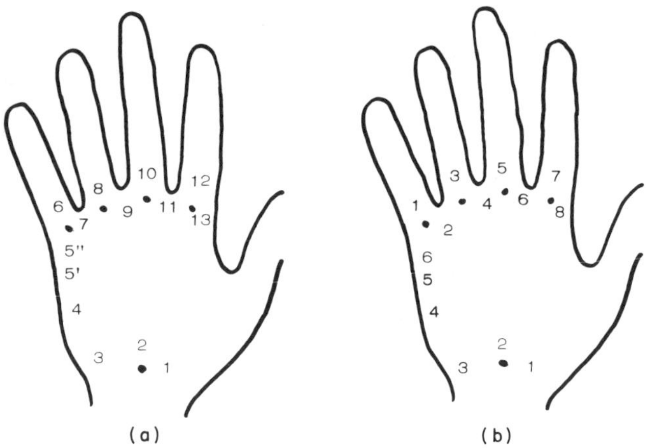

Figure 7.7 (a) Numbers employed in formulating the main lines, and (b) numbers employed in formulating the values of the main-line index.

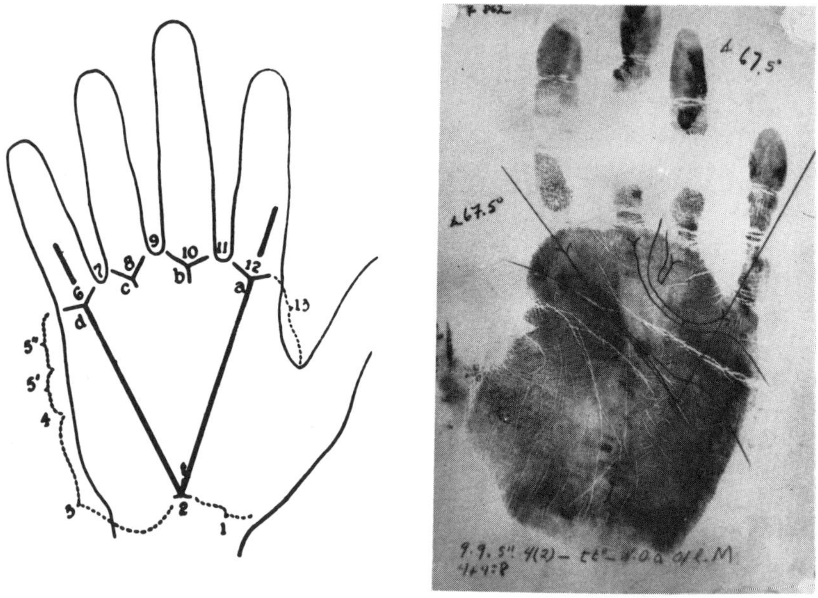

Figure 7.8 (a) Measurement of the *atd* angle in a normal position; and (b) *atd* angle of a women athlete showing the *t''* position 67.5.

5. *The Maximal atd Angle*

The measurement of the *atd* angle was first proposed by Penrose (1949, 1954). The maximal angle is that formed by two straight lines from the most distal axial triradius *t* to the most lateral palmar triradius *a* and the most medial *d* (Figure 7.8). The *atd* value for an individual is the sum of the angles on right and left hands.

III. Inheritance

Although Galton was the first investigator to study the question of heredity in twins, fraternal couplets, and parents and children, no definite results could be drawn from his studies. In the first decade of this century, Locard, Senet, and Stockies denied that finger-print patterns were transmitted by heredity, since their observations of children and parents did not reveal identical dermatoglyphic patterns. In 1939, Böhmer and Harren concluded from their studies that inheritance was not playing a role in determining pattern types. Now there is evidence that some dermatoglyphic traits are inherited. Until recent years, genetic studies have been concerned with dermatoglyphic traits on fingers and palms.

Attempts to determine the mode of inheritance by means of qualitative studies have given inconclusive results. The inheritance of "twisting" i.e. double core patterns, has been variously ascribed to a dominant gene (Bonnevie, 1924), an incompletely dominant gene (Muller, 1930) and to a recessive gene (Karl, 1934). In no case is the genetic situation satisfactorily explained by a single gene factor inheritance. In order to learn more about the inheritance of dermatoglyphic traits quantitative methods and statistical analysis have been used. From familial data, correlations between parents and children, between siblings, and between twins have been calculated for various measurements, chiefly on fingers and palms (see Table 7.1).

Assuming random mating and additive gene effects, the theoretical value of the parent–child, sib–sib, and dizygotic twin correlation coefficients is 0.50, that for monozygotic twins is 1.00. For total finger ridge count, the theoretical and the observed values agree well, as seen in Table 7.1.

The genetics of dermatoglyphic traits is complicated. The ridges are formed during the third and fourth fetal months. Growth disturbances at this time, whether the cause is genetic or environmental, affect the arrangement of the ridges. At present we know little of how these agents act.

TABLE 7.1. CORRELATION COEFFICIENTS BETWEEN RELATIVES FOR SOME DERMATOGLYPHIC TRAITS ACCORDING TO PENROSE (1966)

	Total Finger Ridge count (Holt, 1961)	Measurement of atd angle (Penrose, 1963 b)	Third Inter-digital pattern (Fang, 1950)
Parent-child	r = 0.49 ± 0.04	0.30 ± 0.04	0.43 ± 0.06
Mother-child	r = 0.49 ± 0.04		0.26 ± 0.07
Father-child	r = 0.50 ± 0.04		0.32 ± 0.07
Sibs	r = 0.50 ± 0.05	0.37 ± 0.04	0.27 ± 0.06
Dizygotic twins	r = 0.49 ± 0.08		
Monozygotic twins	r = 0.95 ± 0.01	0.63 ± 0.09	0.78 ± 0.08

IV. Material and Methods

Palm and finger prints were taken in the conventional way using printers ink, plain paper, cards, and a rubber roller. A thin film of ink was spread on a plate, and the roller was inked from this. The roller was then applied to the whole surface of the palm including fingers, in such a way as to ensure that there was a thin, even film of ink over the whole area of ridged skin.

The hand was then pressed against a 5 inch by 8 inch card, taking care that the center of the palm made contact with the paper. Finger tips were inked by rolling each one separately on the ink on the plate, and rolled one by one on a 3 inch by 5 inch card so that the complete pattern of each was recorded. It is essential that prints are complete and not smudged. In order to get complete prints of all finger patterns, each finger must be rolled on the recording surface. Before printing, it is necessary to clean and dry the skin. Sweaty hands and the use of an excessive amount of ink cause blurring and make it impossible to obtain clear prints. Interpretation and formulation of the material was done according to the methods described by Cummins and Midlo (1961).

The group reported on here represents 1227 athletes. Prints were obtained from 1081 males and 146 females from 92 countries. Specialists in 12 sports were included. In view of the small number in each athletic event and the division into groups (i.e., track and field: 100 meters, 800 meters; javelin; weight throwing; etc.) and the large number of races involved, all athletes were grouped into a broad ethnic classification: Caucasoids, Negroids, Mongoloids, and Mestizos; and a physiological classification: explosive, speed, skill, and endurance.

This physiological classification consists of four groups: the sports (*1*) where energy is released in a few seconds (explosive), (*2*) where speed is essential (speed), (*3*) where the main characteristic is the need for specialized technique (skill), and (*4*) where efficiency depends on the resistance of the individual (endurance) (see Table 7.2).

TABLE 7.2. NUMBER OF INDIVIDUALS IN EACH GROUP

	Caucasoids	Negroids	Mongoloids	Mestizos
Males				
Explosive	112	63	23	46
Speed	64	81	11	25
Skill	144	28	20	38
Endurance	289	31	29	77
Females				
Explosive	14	7	0	4
Speed	28	19	1	11
Skill	24	5	0	9
Endurance	19	0	2	3

The following dermatoglyphic features were analyzed:

1. Finger prints: pattern types (arches, ulnar and radial loops, whorls); ridge count; pattern index intensity.
2. Palm prints: main line terminations; main line index; position of axial triradii; *atd* angle; configuration types in respect to the five palmar areas: hypothenar, thenar/interdigital I, interdigitals II, III, and IV.

Only the following traits are taken as general parameters to study the athletic population:

1. *Finger prints:* pattern types (arches, loops, and whorls); pattern size (total ridge count); pattern index intensity.
2. *Palm prints:* main-line index; absence of main line C; *atd* angle.

V. Results and Discussion

The results for finger prints are given in Tables 7.3, 7.4, 7.5, and Figure 7.9, and for palm prints in Tables 7.6, 7.7, 7.8, and Figure 7.10.

The study of dermatoglyphics offers the possibility of finding evidence that characters are genetically determined. The present study was undertaken in the hope that evidence would be provided that athletes of different kinds show inherited differences in their manual dermatoglyphics. Unfortunately, the racial classification is too broad to be satisfactory for dermatoglyphic purposes. We know that differences exist between the populations of western European countries. Much greater differences are apparent in the dermatoglyphics of Asian countries. Because of differences in the methods of classification, it is not possible to compare results directly with those of other authors. Therefore, data gathered from published sources have been selected and averaged to provide rough comparisons. The many subdivisions involved result in the number of observations in some of the groups being rather small. Hence in various cases the apparent differences are not statistically significant. However, it is valuable to have these data available so that they can be combined with the results of future studies of this type. They may then contribute to the discovery of any significant differences in dermatoglyphic patterns that may exist among the athletes in the various sport specialities.

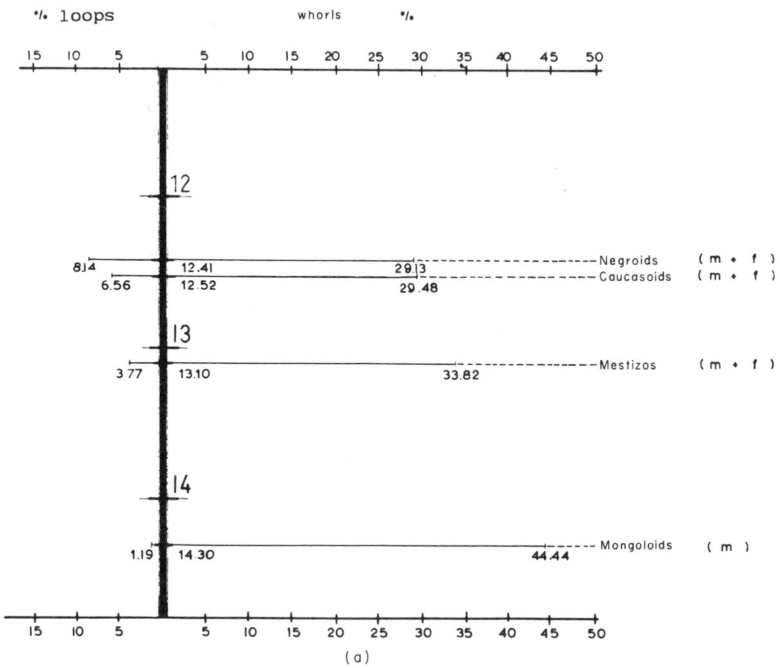

Figure 7.9 Index of pattern intensity according to races and sport groups: (a) athletic population according to races; (b) Caucasoids; (c) Negroids; (d) Mongoloids; and (e) Mestizos. (Continued on the following two pages.)

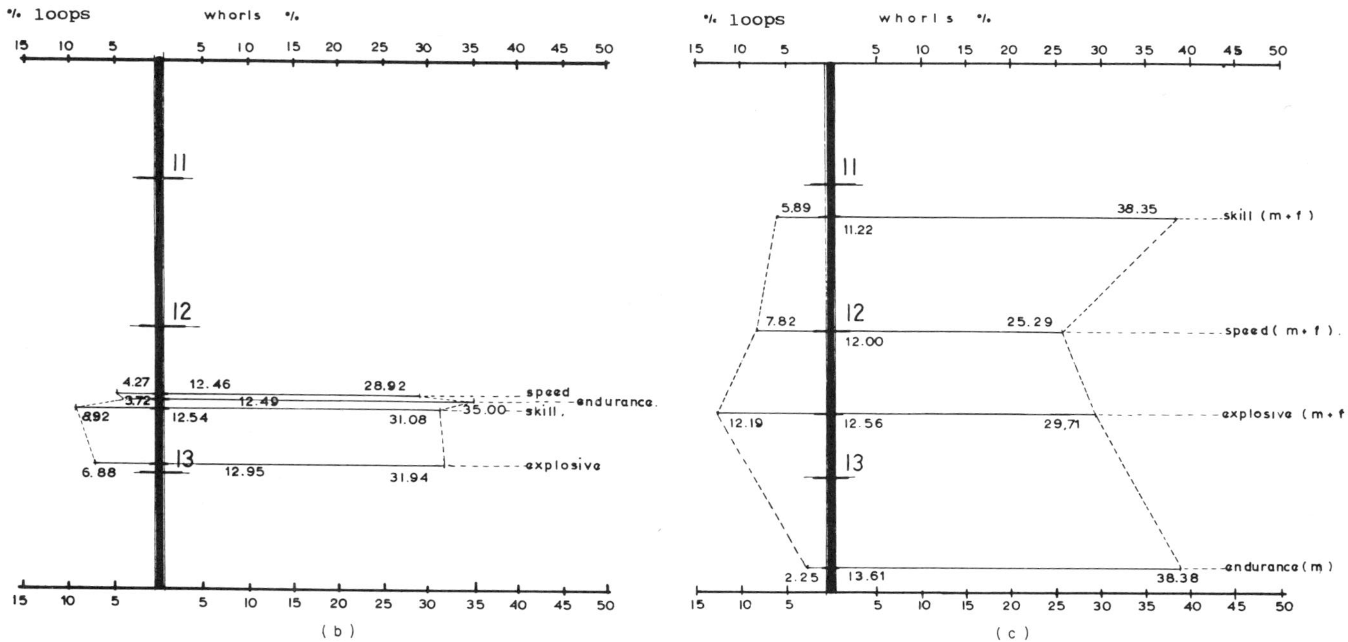

Figure 7.9 (continued).

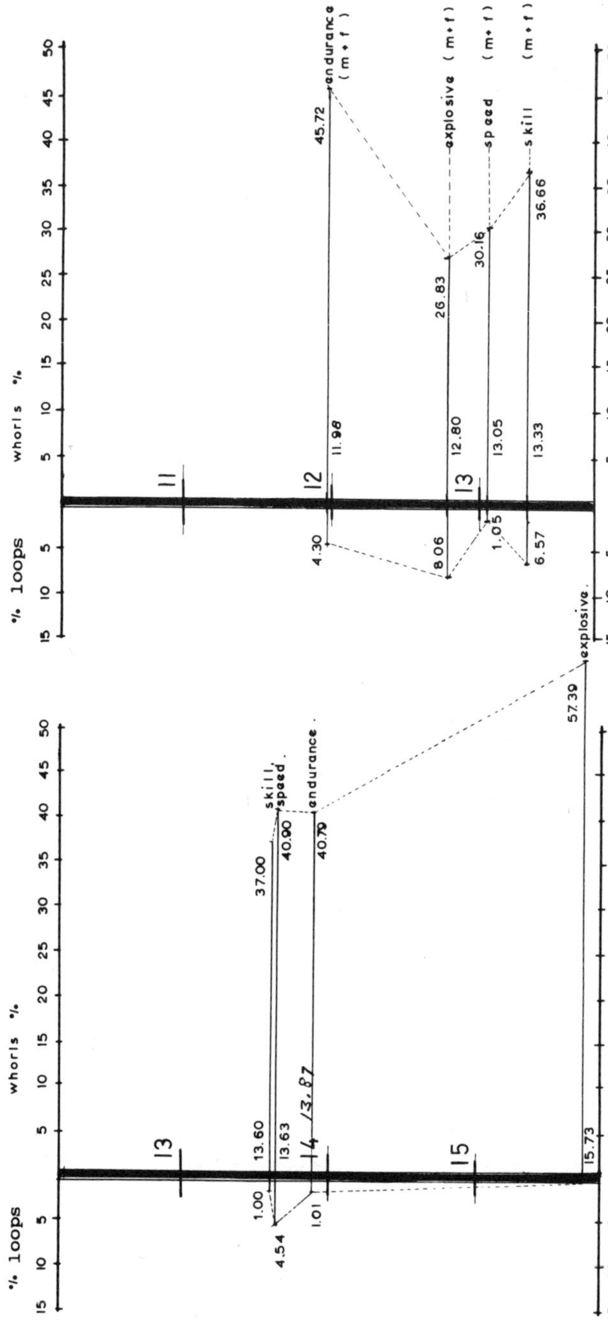

Figure 7.9 (continued).

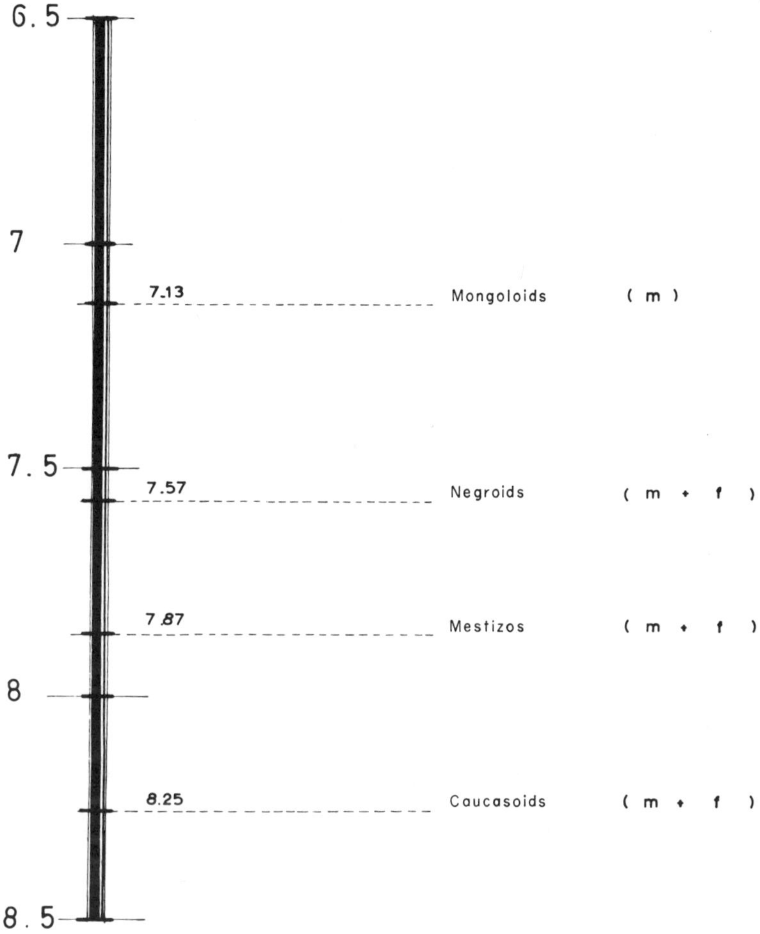

Figure 7.10 Main-line index according to races.

TABLE 7.3. PERCENTAGE OF PATTERN TYPES ON FINGERS OF ATHLETES ACCORDING TO RACE

	CAUCASOIDS (85 female, 609 male)									NEGROIDS (31 female, 203 male)								
	RIGHT			LEFT			BOTH			RIGHT			LEFT			BOTH		
	A	L	W	A	L	W	A	L	W	A	L	W	A	L	W	A	L	W
FEMALES																		
Explosive	5.7	61.4	32.9	5.7	72.9	21.4	5.7	67.7	27.1	14.3	62.8	22.8	20.0	54.3	25.7	17.1	58.6	24.3
Speed	5.0	65.7	29.3	4.3	67.1	28.6	4.6	66.4	28.9	7.4	72.6	20.0	9.5	68.4	22.1	8.4	70.5	21.0
Skill	11.7	57.5	30.8	13.3	58.3	28.3	12.5	57.9	29.6	0.0	48.0	52.0	5.0	45.0	50.0	2.2	46.7	51.7
Endurance	6.3	74.7	18.9	10.5	70.5	18.9	8.4	72.6	18.9									
Total	7.3	64.7	28.0	8.5	66.3	25.2	4.5	37.8	57.7	7.7	66.4	25.8	11.3	62.0	26.7	9.5	64.3	26.2
MALES																		
Explosive	6.5	51.7	41.7	5.6	62.7	31.6	6.1	57.2	36.7	7.1	56.9	36.0	7.4	58.3	34.3	7.2	57.6	35.1
Speed	3.7	61.6	34.7	4.1	72.7	23.2	3.9	67.1	28.9	6.6	61.6	31.8	7.8	65.3	26.9	7.2	63.4	29.3
Skill	4.9	59.0	36.1	5.8	65.6	28.6	5.3	62.3	32.4	8.6	60.7	30.7	10.0	69.3	20.7	9.3	65.0	25.7
Endurance	5.6	58.5	35.9	4.8	67.9	27.3	5.2	63.2	31.6	1.9	53.8	43.2	2.6	63.9	33.5	2.3	59.4	38.3
Total	5.4	57.7	36.9	5.1	66.9	28.0	5.3	62.3	32.5	6.3	59.0	34.7	7.2	63.4	29.4	6.8	61.2	32.0

AVERAGE VALUES* OF VARIOUS AUTHORS:

Caucasoids:
- Dutch (m) - Dankmeijer, 1934
- English - Galton, 1892
- Germans (m) - Eydt, 1933
- Hungarians - Bonnevie, 1929
- Italians - Falco, 1908
- Norwegians - Bonnevie, 1924
- Portuguese - Lopes, 1931
- Russians (m) - Smenoysx, 1927

A	L	W
5.8	63.4	30.8

Negroids (Sierra Leone) - Dankmeijer, 1938
Mulattos (Jamaica) - Davenport and Steggerda, 1929

A	L	W
7.9	65.3	26.6

	MONGOLOIDS 83 male									MESTIZOS 27 female 186 male								
	RIGHT			LEFT			BOTH			RIGHT			LEFT			BOTH		
	A	L	W	A	L	W	A	L	W	A	L	W	A	L	W	A	L	W
FEMALES																		
Explosive										10.0	55.0	35.0	15.0	80.0	5.0	12.5	67.5	20.0
Speed										1.8	69.1	29.1	1.8	61.8	36.4	1.8	65.4	32.7
Skill										0.0	64.4	35.5	0.0	68.9	31.1	0.0	66.7	33.3
Endurance										6.7	60.0	33.3	0.0	66.7	33.3	3.3	63.3	33.3
Total										3.0	64.4	32.6	3.0	67.4	29.6	2.9	65.9	31.1
MALES																		
Explosive	0.0	39.1	60.9	0.0	46.1	53.9	0.0	42.6	57.3	4.2	59.6	36.2	3.0	65.8	31.2	3.6	62.7	33.7
Speed	3.6	52.7	43.6	5.4	56.4	38.2	4.5	54.5	40.1	2.4	64.8	32.8	0.0	69.6	30.4	1.2	67.2	31.6
Skill	0.0	65.0	35.0	2.0	59.0	39.0	1.0	62.0	37.0	6.8	47.9	45.3	6.3	58.9	34.7	6.6	53.4	40.0
Endurance	1.3	56.4	42.3	6.7	60.0	39.3	1.0	58.2	40.8	4.3	55.9	39.8	6.3	57.3	36.4	5.3	56.6	38.1
Total	0.9	53.2	45.8	1.4	55.5	43.1	1.2	54.3	44.5	4.5	56.4	39.1	4.6	61.3	34.0	4.6	58.9	36.5

AVERAGE VALUES* OF VARIOUS AUTHORS:	Sources	A	L	W
Mongoloids	Chinese - Kubo, 1918; Japanese (m) - Hirano, 1931; Japanese - Kubo, 1918; Koreans - Kubo, 1918	1.7	50.3	47.8
Mestizos	South Americans Chile - Henckel, 1933; 61,545 individuals	4.8	58.9	36.3
	URBAN POPULATION (MEXICO) Cobo de Gallegos 1964; 924 individuals	6.1	62.2	31.7.

* Collected figures from Dankmeijer, Am. Jour. Hum. Genet., 1938.
The separate figures of these groups are to be found in his publications.

TABLE 7.4. AVERAGE NUMBER OF TRIRADII PER INDIVIDUAL (PATTERN INTENSITY INDEX), ACCORDING TO RACES AND SPORTS GROUPS

	Caucasoids	Negroids	Mongoloids	Mestizos
Explosive	12.95	12.56	15.73	12.80
Speed	12.46	12.00	13.63	13.05
Skill	12.54	11.22	13.60	13.33
Endurance	12.49	13.61	13.87	11.98
Total athlete population	12.52	12.41	14.30	13.10

TABLE 7.5. TOTAL RIDGE-COUNT AND S.D. OF THE ATHLETES ACCORDING TO RACE AND SPORT GROUP

	CAUCASOIDS			NEGROIDS		
	R	L	B	R	L	B
MALES						
Explosive	70.93 ± 24.17	69.28 ± 24.78	140.36 ± 47.53	62.63 ± 26.32	60.92 ± 26.22	123.62 ± 52.14
Speed	66.07 ± 22.27	64.93 ± 22.70	131.23 ± 43.55	60.25 ± 20.31	56.71 ± 23.29	117.10 ± 42.45
Skill	66.97 ± 24.63	65.16 ± 25.61	133.13 ± 49.68	57.89 ± 22.63	55.85 ± 23.12	113.75 ± 45.04
Endurance	67.86 ± 24.12	65.26 ± 23.56	133.12 ± 46.73	65.80 ± 21.17	65.61 ± 23.28	131.41 ± 44.03
FEMALES						
Explosive	68.85 ± 19.52	60.07 ± 25.85	128.92 ± 44.32	45.00 ± 29.04	40.71 ± 28.94	85.71 ± 57.81
Speed	55.96 ± 20.15	54.89 ± 21.26	110.85 ± 40.55	60.05 ± 26.71	57.73 ± 29.02	112.52 ± 55.08
Skill	62.50 ± 33.27	60.54 ± 33.51	123.04 ± 66.13	91.00 ± 9.30	81.00 ± 19.79	170.00 ± 28.74
Endurance	57.78 ± 21.80	55.52 ± 21.19	113.84 ± 42.15			

	MONGOLOIDS			MESTIZOS		
	R	L	B	R	L	B
MALES						
Explosive	75.95 ± 14.93	73.86 ± 17.89	145.47 ± 31.27	65.50 ± 25.00	63.00 ± 21.58	129.51 + 44.58
Speed	64.72 ± 20.46	64.72 ± 22.33	129.45 ± 42.46	67.80 ± 20.41	66.60 ± 19.42	134.40 ± 39.03
Skill	69.50 ± 19.20	67.90 ± 19.77	137.40 ± 37.72	69.28 ± 24.33	66.23 ± 23.34	133.32 ± 47.10
Endurance	70.31 ± 17.08	67.30 ± 18.20	137.55 ± 34.91	67.63 ± 26.50	65.00 ± 28.22	134.43 ± 54.12
FEMALES						
Explosive				51.25 ± 26.51	44.50 ± 25.00	95.75 ± 50.92
Speed				69.63 ± 18.83	62.27 ± 24.61	131.90 ± 43.94
Skill				76.66 ± 11.06	69.11 ± 9.43	145.77 ± 18.53
Endurance						

TABLE 7.6. PERCENTAGE AND S.D. OF ABSENCE OF MAIN LINE C AND ITS ABORTIVE STATES X, x, 0

	CAUCASOIDS			NEGROIDS		
	R	L	B	R	L	B
MALES						
Explosive	2.54 ± 0.16	17.80 ± 0.35	10.17 ± 0.38	6.15 ± 0.23	9.23 ± 0.28	7.69 ± 0.33
Speed	4.69 ± 0.21	15.63 ± 0.34	10.16 ± 0.40	8.54 ± 0.27	21.95 ± 0.37	15.24 ± 0.41
Skill	10.96 ± 0.30	15.75 ± 0.33	13.36 ± 0.44	14.29 ± 0.33	17.86 ± 0.35	16.07 ± 0.47
Endurance	9.52 ± 0.27	15.65 ± 0.33	12.59 ± 0.45	12.90 ± 0.32	19.35 ± 0.36	16.13 ± 0.53
FEMALES						
Explosive	7.14 ± 0.26	14.29 ± 0.34	10.71 ± 0.38	0.00 ± 0.00	14.29 ± 0.35	7.14 ± 0.35
Speed	7.14 ± 0.25	25.00 ± 0.38	16.07 ± 0.47	0.00 ± 0.00	5.26 ± 0.22	2.63 ± 0.22
Skill	4.35 ± 0.20	13.04 ± 0.32	8.70 ± 0.35	0.00 ± 0.00	20.00 ± 0.40	10.00 ± 0.40
Endurance	15.79 ± 0.34	36.84 ± 0.39	26.32 ± 0.64			

	MONGOLOIDS			MESTIZOS		
	R	L	B	R	L	B
MALES						
Explosive	4.35 ± 0.21	8.70 ± 0.27	6.52 ± 0.32	23.40 ± 0.37	25.32 ± 0.38	24.47 ± 0.60
Speed	18.18 ± 0.36	9.09 ± 0.28	13.64 ± 0.40	16.00 ± 0.34	24.00 ± 0.38	20.00 ± 0.39
Skill	20.00 ± 0.37	15.00 ± 0.34	17.50 ± 0.50	18.42 ± 0.35	18.42 ± 0.38	18.42 ± 0.50
Endurance	3.45 ± 0.18	6.67 ± 0.24	5.06 ± 0.39	7.50 ± 0.25	10.00 ± 0.29	8.75 ± 0.38
FEMALES						
Explosive				0.00 ± 0.00	25.00 ± 0.43	12.50 ± 0.43
Speed				27.27 ± 0.40	18.18 ± 0.36	22.73 ± 0.57
Skill				0.00 ± 0.00	22.22 ± 0.39	11.11 ± 0.39
Endurance				0.00 ± 0.00	33.33 ± 0.47	16.67 ± 0.47

TABLE 7.7. MAIN-LINE INDEX AND S.D. ACCORDING TO RACES AND SPORT GROUPS

	CAUCASOIDS			NEGROIDS		
	R	L	B	R	L	B
MALES						
Explosive	9.23 ± 1.87	7.72 ± 2.38	8.48 ± 2.26	8.73 ± 1.18	6.73 ± 2.17	7.73 ± 2.25
Speed	8.82 ± 2.24	7.07 ± 2.28	7.95 ± 2.42	8.28 ± 2.23	6.91 ± 2.16	7.59 ± 2.29
Skill	9.12 ± 2.07	7.50 ± 1.99	8.31 ± 2.17	7.96 ± 2.38	7.28 ± 2.41	7.62 ± 2.46
Endurance	9.04 ± 1.94	7.49 ± 2.08	8.27 ± 2.15	8.22 ± 2.18	6.83 ± 2.17	7.53 ± 2.25
Total	9.08 ± 1.98	7.50 ± 2.15	8.29 ± 2.21	8.37 ± 2.14	6.90 ± 2.19	7.65 ± 2.28
FEMALES						
Explosive	8.00 ± 2.03	7.14 ± 2.34	7.57 ± 2.20	7.42 ± 1.71	6.85 ± 1.57	7.14 ± 1.60
Speed	9.10 ± 2.11	7.14 ± 2.50	8.12 ± 2.50	7.94 ± 2.23	6.26 ± 1.99	7.10 ± 2.24
Skill	8.65 ± 2.03	7.43 ± 2.59	8.04 ± 2.38	7.60 ± 3.20	7.40 ± 3.36	7.50 ± 3.10
Endurance	8.73 ± 1.99	7.15 ± 2.19	7.94 ± 2.21			
Total	8.71 ± 2.04	7.23 ± 2.39	7.97 ± 2.34	7.77 ± 2.23	6.58 ± 2.14	7.17 ± 2.24
Total Values and S.D. for males and females according to race	9.05 ± 1.99	7.47 ± 2.18	8.25 ± 2.23	8.29 ± 2.16	6.85 ± 2.18	7.57 ± 2.28

	MONGOLOIDS			MESTIZOS		
	R	L	B	R	L	B
MALES						
Explosive	7.30 ± 1.96	5.95 ± 2.16	6.63 ± 2.15	8.76 ± 1.97	6.95 ± 1.98	7.86 ± 2.17
Speed	9.00 ± 1.41	7.09 ± 2.21	8.04 ± 2.05	8.84 ± 1.88	7.16 ± 2.15	8.00 ± 2.17
Skill	7.95 ± 2.06	6.35 ± 2.05	7.15 ± 2.18	9.23 ± 1.86	7.36 ± 2.10	8.30 ± 2.19
Endurance	8.10 ± 2.19	6.30 ± 2.08	7.18 ± 2.30	8.75 ± 2.07	6.80 ± 1.98	7.77 ± 2.24
Total	7.96 ± 2.04	6.32 ± 2.10	7.13 ± 2.22	8.86 ± 1.97	7.00 ± 2.02	7.93 ± 2.20
FEMALES						
Explosive				9.25 ± 3.50	7.25 ± 2.87	8.25 ± 3.14
Speed				8.54 ± 2.62	6.63 ± 2.69	7.59 ± 2.77
Skill				8.33 ± 2.39	6.77 ± 2.22	7.55 ± 2.38
Endurance				8.71 ± 1.70	8.00 ± 1.28	8.35 ± 1.49
Total				8.61 ± 2.38	6.45 ± 1.87	7.53 ± 2.39
Total Values and S.D. for males and females according to race				8.83 ± 2.03	6.92 ± 2.01	7.87 ± 2.14

TABLE 7.8. atd ANGLE IN DEGREES AND S.D. IN ATHLETES ACCORDING TO RACES AND SPORT GROUPS

Values on Both hands	CAUCASOIDS SUM of R + L	NEGROIDS SUM of R + L	MONGOLOIDS SUM of R + L	MESTIZOS SUM of R + L
Males				
Explosive	87.53 ± 10.46	82.70 ± 7.32	78.86 ± 3.18	89.43 ± 8.31
Speed	86.59 ± 7.84	77.56 ± 8.16	80.08 ± 7.19	84.47 ± 6.50
Skill	89.04 ± 9.74	79.99 ± 7.39	84.72 ± 6.30	82.61 ± 6.78
Endurance	84.61 ± 10.08	80.17 ± 7.75	79.50 ± 5.20	82.41 ± 7.71
Females				
Explosive	78.67 ± 9.49	83.52 ± 10.62		85.62 ± 5.26
Speed	82.31 ± 8.91	83.99 ± 8.08		76.31 ± 18.15
Skill	82.58 ± 4.21	86.20 ± 7.89		83.17 ± 17.22
Endurance	92.69 ± 10.59			80.09 ± 2.57

Bibliography

Bidloo, G. (1685). "Anatomia humani corporis." Amsterdam. (Utrecht edition, 1728.)

Bohmer, K., and Harren, F. (1939). Die Vererbung der Papillar linien and ihre Bedeutung fur den Nachweis der Vaterschaft. *Deutsche Ztschr. f.d. ges. gerichtl. Med.* **32,** 73–82.

Bonnevie, K. (1924). Studies on papillary patterns of human fingers. *J. Genet.* **15,** 1.

Cummins, H. (1936). Methodology in palmar dermatoglyphics. *Middle Am. Research Series.* Tulane Univ. 7.

Cummins, H., and Midlo, C. (1961). Finger prints, palms and soles." Dover Publications, Inc., New York.

Cummins, M., and Steggerda, M. (1935). Finger prints in a dutch family series. *Am. J. Phys. Anthrop.* **XX** (No. 1) and supplement. April–June.

Dankmeijer, J. (1938). Dermatoglyphics with some anthropological data. *Am. J. Phys. Anthrop.* **23,** 377–388.

Fang, T. C. (1950). The third interdigital patterns on the palms of the general British population mongoloid and nonmongoloid mental defectives. *J. Ment. Sci.* **96,** 780.

Faulds, H. (1880). "Finger-print identification." (Cited by Cummins, H., and Midlo, C. "Finger prints, palms and soles." Dover Publications, Inc., New York, 1961.)

Galton, F. (1892). "Finger-prints." Macmillan and Co., London.

Galton, F. (1895). "Fingerprint directories." Macmillan and Co., London.

Grew, H. (1684). (Cited by Wentworth, B., and Wilder, H. H. "Personal identification," 2nd ed. T. G. Cooke, Chicago, 1932.)

Henry, E. R. (1901). "Classification and uses of finger prints." (8th ed. published 1937 by H. M. Stationary Office, London.)

Holt, S. B. (1961a). "Dermatoglyphic patterns." Pergamon Press, New York.

Holt, S. B. (1961b). Quantitative genetics of finger print patterns. *Brit. Med. Bull.* **17**, 247.

Karl, E. (1934). "Systematische und erbbiologische Untersuchungen der Papillarmuster der menschlichen Fingerbeeren." Dissertation Leipzig. (See Cummins and Midlo, 1943, "Finger prints, palms and soles.")

Malpighi, M. (1686). "De externo tactus organo." London.

Muller, B. (1930). "Untersuchungen über die Erblichkeit von Fingerbeerenmustern unter besonderer Berücksichtingung rechtlicher Fragestellungen." *Z. indukt. Abstamm. Vererb.* Lehre **56**, 302.

Penrose, L. S. (1949). Familial studies on palmar patterns in relation to mongolism. *Hereditas*, Supp. Vol., 412.

Penrose, L. S. (1954). The distal triradius t on the hands of parents and sibs of mongol imbeciles. *Ann. Hum. Genet.* **19**, 10.

Penrose, L. S. (1963). "Biology of mental defect," 3rd. ed. Sedwick and Jackson, Ltd., London.

Penrose, L. S., and Smith, G. F. (1966). "Down's anomaly." Churchill, Ltd., London.

Purkinje, J. E. (1940). "Commentatio de examine physiologico organi visus et systematis cutanei." Breslau, 1823. (Translated into English by Cummins, H., and Kennedy, R. W. *Am. J. Crim. Law and Criminol.* **31**, 343–356.)

8

Cytogenetics

I. Cytological Investigation

A. *Sex Chromatin Analysis*

The term "sex chromatin" denotes a body in the cell nuclei that has a close connection with the sex of the person from whom the cells were taken. In the large majority of human beings, and the same applies to most mammals, the presence of a sex chromatin body in a portion of cell nuclei indicates that the donor of the cells was a female, while the absence of sex chromatin bodies indicates that the cells originated from a male.

The discovery of sex chromatin dates from the year 1949, when Barr and Bertram, in London, Canada, found that a specific body was present in the neurones of female cats and was absent from the neurones of male cats. For this reason the sex-chromatin body of tissue cells is often called a "Barr body." This term has the advantage of brevity and is also free from any emotional overtones.

Following the discovery in the cat, Barr bodies were found in human females, first in nerve cells (Barr *et al.*, 1950) and then in sections obtained from skin biopsies (Moore *et al.*, 1953). The findings that Barr bodies could be demonstrated in cells from the buccal mucosa opened up a wide field of research. For this technique it is only necessary to scrape the inside of the cheeks with a spatula and the cells so obtained are spread on a glass slide and prepared for microscopy (Moore and Barr, 1955; Marberger *et al.*, 1955). The ease of the method and the complete lack of discomfort to the person to be tested has made it possible to use it on very large numbers of people, including newborn babies (Maclean, 1966), patients suspected of abnormalities of sexual development (Barr, 1966), and patients in hospitals for the mentally retarded as well as in mental hospitals (Maclean *et al.*, 1962, 1968).

Barr bodies may be stained with any type of dye used for staining cell nuclei and chromosomes. The best results are probably obtained by using the Feulgen reaction, which is specific for the presence of deoxyribonucleic acid (DNA). The reagent used in this reaction is a decolorized solution of the dye fuchsin. The cells to be stained are subjected to mild acid hydrolysis, which liberates aldehydes from the sugar component of the DNA; these aldehydes restore the color of the fuchsin (Pearse, 1968). As a result, nuclei and their component structure assume a purple color, while the cytoplasm remains unstained. The Barr bodies are dense, deeply stained structures with a diameter of about one micrometer (pm). Typical Barr bodies are semicircular or triangular in shape, with the pointed end directed towards the center of the nucleus. The majority are situated against the periphery of the cell nucleus (Figure 8.1a and b). When examining cells from the buccal mucosa, it is usual to accept as Barr bodies only those structures, of appropriate size and shape, that lie directly against the nuclear membrane. Even though this procedure underestimates the true incidence of Barr bodies, it avoids confusion of the true Barr bodies with other nuclear aggregates that are present in some nuclei. These nonspecific structures are usually known as "chromocenters." A number of these may be present in a single nucleus (Figure 8.2).

Instead of the Barr bodies of tissue cells, the white blood cells known as polymorphonuclear leucocytes carry a specific nuclear appendage known as a "drumstick" (Davidson and Smith, 1954). A drumstick consists of a head of about 1.5 micrometers in diameter, which is attached to the rest of the nucleus by a filament of varying length (Figure 8.3). Drumsticks, like Barr bodies, occur in normal females and are absent in normal males. However, the frequency with which drumsticks occur is much lower than that of Barr bodies. The incidence of drumsticks

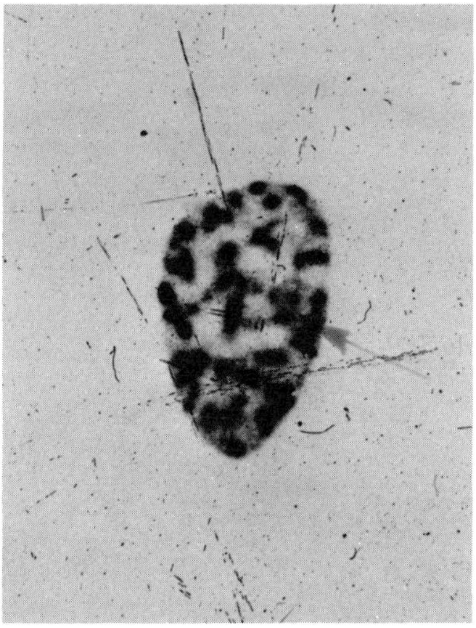

(a)

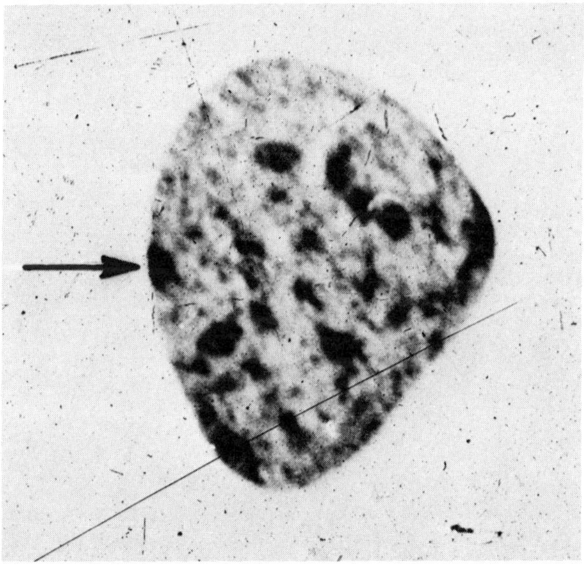

(b)

Figure 8.1 (a) Buccal mucosal cell from a male; (b) Buccal mucosal cell from a female, showing Barr body.

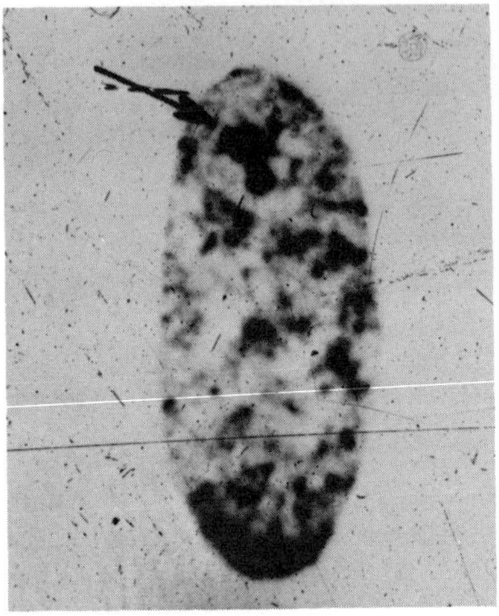

Figure 8.2 Cell showing "chromocenter."

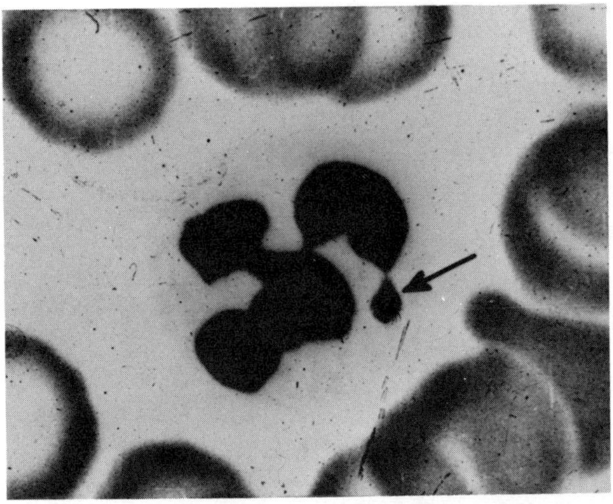

Figure 8.3 White blood cell from female showing "drumstick."

is only about 3 per 100 polymorphonuclear neutrophil leucocytes of normal females (Davidson, 1966; Mittwoch, 1967). The enumeration of drumsticks, is, therefore, rather time consuming and it was decided that the drumstick test should not be carried out on the athletes.

The chromosomes can be seen only in cells which are either in mitosis or in meiosis. By contrast, Barr bodies are present only in nondividing cells, the nuclei of which are usually described as resting, or in interphase.

In normal females, one Barr body is present in a substantial proportion of cells, while Barr bodies are absent in cells from males. This is connected with the fact that if two X-chromosomes are present, one of them behaves differently from the first, and from the rest of the chromosomes.

Although the chromosomes are visible only during cell division, chromosome duplication takes place during the interphase prior to this division. One of the necessary raw materials for chromosome replication is thymidine and by supplying the cells with a radioactive form of thymidine (tritiated thymidine) it is possible to obtain an idea of the time when chromosome replication takes place. When this technique is applied to cells from females, it will be seen that one chromosome replicates later than the rest of the chromosomes (German, 1964). There is a great deal of evidence that this chromosome is in fact an X-chromosome. A similar late replicating chromosome is not found in cells from normal males. However, in the cells of males with Klinefelter's syndrome and with sex-chromosome constitution XXY, a late replicating chromosome resembling an X-chromosome is present and the cells of such patients contain a Barr body in a substantial proportion of their nuclei.

Moreover, in patients with three X-chromosomes, both females of constitution 47,XXX, and males with constitution 48,XXXY, two late replication X-chromosomes can be seen in the dividing cells and two Barr bodies in interphase cells (Figure 8.4). If four X-chromosomes

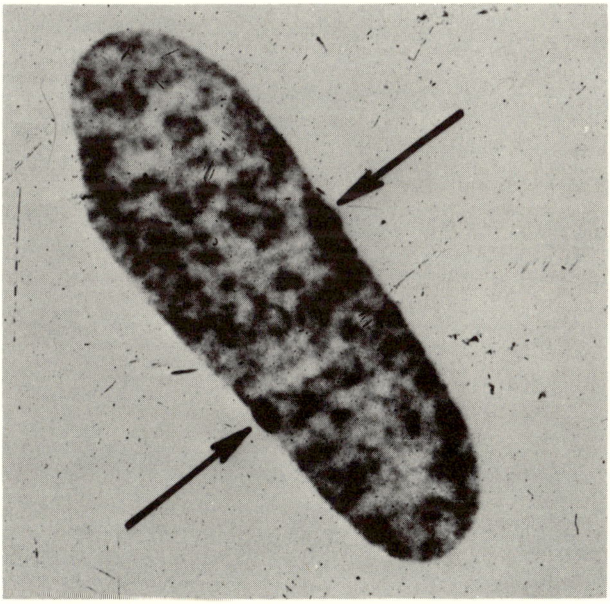

Figure 8.4 Two Barr bodies as found in cells containing three X-chromosomes.

are present in the karyotype, three late replicating X-chromosomes can be seen in dividing cells, and three Barr bodies in some of the interphase nuclei (see Mittwoch, 1967). Lastly, in exceptional female patients with five X-chromosomes, four late replicating X-chromosomes and four Barr bodies can be seen in some of the cells (Grumbach *et al.*, 1963; Ricci *et al.*, 1968).

It is clear that the maximum number of Barr bodies in nondividing cells corresponds to the maximum number of late replicating X-chromosomes seen in dividing cells and that this number is one less than the total number of X-chromosomes, present in the karyotype. This shows that when more than one X-chromosome is present, the additional X-chromosome or X-chromosomes behave differently from the rest of the chromosomes in at least two ways: (*1*) they replicate their DNA late and (2) they are highly condensed in interphase nuclei when the rest of the chromosomes are largely extended. Chromosomes which differ in their DNA cycle as well as their staining reaction from the rest of the chromosomes are called "allocyclic" or "heterochromatic" (Darlington and La Cour, 1940; Heitz, 1928). Therefore, any X-chromosome apart from the first one present in the karyotype assumes the properties of heterochromatin. It is likely that the multiple chromocenters which are seen in some interphase cells represent heterochromatic regions of autosomes.

According to classical genetic theory it was assumed that heterochromatin carries few, if any, Mendelian genes (Muller and Painter, 1932; Heitz, 1933; Mather, 1944; Commoner, 1964). This idea was largely based on findings in *Drosophila.* In recent years the idea of the relative genetic inertness of heterochromatin has been applied by Lyon (1961; 1966) to the heterochromatic X-chromosome or X-chromosomes in man and other mammals. According to this hypothesis, the genes carried on the X-chromosome which forms the Barr body are not expressed. This would mean that each cell in a female carries only one X-chromosome with active gene expression, just like the cells from males. Nevertheless, females differ from males because in some of their cells the paternal X-chromosome is actively expressing its genes, while in other cells the maternal X-chromosome is active in this way. The Lyon hypothesis has been, and continues to be, a subject of active discussion and has stimulated a large number of experimental investigations.

It will be seen that the presence of Barr bodies, as well as their numbers in individual nuclei, gives an indication of the number of X-chromosomes present. This will be one more than the maximum number of Barr bodies present per cell. On the other hand this test will give no information of the presence of Y-chromosomes. Thus, a person in whose cells no Barr bodies are present may be of sex chromosome constitution 46, XY; 47, XYY, or 45, X. Among persons with the 47, XYY type of chromosome constitution, males who are taller than average and who tend to have aggressive tendencies have been found (Jacobs *et al.* 1965; Court-Brown *et al.*, 1968). Even though Goodman *et al.* (1968) failed to find this type of sex chromosome constitution among a group of basketball players, it would still be of interest to see whether it might occur among some athletes and it was decided that chromosome studies should be undertaken of a large number of male Olympic athletes.

The sex chromosome constitution XY, although that of a normal male, has been found in some exceptional females, particularly those with the syndrome of pure gonadal dysgenesis and with testicular feminization (Federman, 1967). This chromosome constitution occurs also in some, though apparently a minority of, persons with true hermaphroditism, who may go through life as either men or women.

Although some women with any of these conditions might be well suited for athletic sports, it was felt that they would probably have been eliminated prior to their arrival in Mexico and that, therefore, it would hardly be worthwhile to undertake chromosome studies on women athletes competing in the Olympic Games. However, since the sex chromatin test from buccal smear requires much less time and effort, it was decided to carry out this test on female athletes.

Although it was the policy to subject male athletes to chromosome analysis, it was thought advisable that the sex chromatin test be carried out on some of them for two reasons. To begin with, a considerable proportion of athletes refused to give blood, which meant that chromosome analysis could not be carried out on them, and the sex chromatin test therefore offered the only possibility of detecting the possible presence of an additional X-chromosome. Second, if any of the men had been mosaics, with sex-chromosome constitution 46,XY in some cells and 47,XXY in other cells, it is conceivable that, if only few cells are available for chromosome analysis, one of the cell lines might be missed; however, the XXY line might show up in the sex chromatin test by the presence of Barr bodies (Barr, 1968).

Mosaicism in sex chromosomes might also occur in women. The phenotype of such women are likely to be variable and would be expected to be influenced by the relative proportion of the different cell lines during particular stages and regions of embryonic development (Klinger and Schwarzacher, 1962). Thus some women with sex-chromosome mosaicism of the type 47,XXX/46,XX; 48,XXXX/46,XX; or 47,XXX/45,X (de Grouchy *et al.*, 1963a; de Grouchy and Lamy, 1963b) might have the phenotype of normal females with physiques that are suitable for athletic sports. The sex chromatin test could not, of course, delineate the entire mosaic karyotype. It might, however, give a clue to the condition by revealing a cell line with multiple sex chromosomes.

Alternatively, it would be useful to exclude the likelihood of the presence of such cell lines in the athletes who were tested.

The meeting of athletes of both sexes belonging to diverse racial groups and different sport specialities led us to assume the possible finding of individuals possessing sex chromosome anomalies associated mainly with some of the extreme anthropological variations which characterized some of the athletes.

B. *Material and Methods*

In this research, samples were taken from 738 athletes of both sexes. The sex chromatin was investigated in buccal smears of 594 male and 144 female athletes. The specimens were taken from athletes of different racial groups (Table 8.1).

TABLE 8.1. DISTRIBUTION OF TESTED ATHLETES BY SEX AND RACE

RACE			M	F
Caucasoid			341	85
	M	F		
(a) Negroid	84	24		
(b) Mulatto	34	6		
(c) Negroid + Mongoloid	1	0		
(d) Mestizo + Negroid	3	1		
Total NEGROID			122	31
Mongoloid			52	2
Mestizo			76	26
Unknown			3	0
TOTAL			594	144

Using a metal spatula, one to three samples were gently scraped from the inside of the mouth of each athlete. The buccal smears were spread on a clean slide, fixed, and stained with the Feulgen technique.

C. *Results and Discussion*

In males the percentage of cells with one Barr body varied from 0% to 3% as follows:

568—0%
19—1%
5—2%
2—3%

In females the percentage of cells with one Barr body varied from 12% to 39% as follows:

2—12%	15—25%
1—13%	9—26%
2—14%	6—27%
2—15%	9—28%
3—16%	2—29%
5—17%	7—30%
9—18%	1—31%
17—19%	1—32%
11—20%	2—33%
16—21%	2—34%
10—22%	1—37%
7—23%	1—39%
13—24%	

In spite of the number of athletes studied, no abnormal case was detected that would suggest a chromosome aberration associated with a sexual syndrome. The possible findings of mosaics involving XXY karyotypes and others were not encountered.

As may be seen there are important variations in the women studied in respect to the percentage of Barr bodies. The average of this percentage is 24% and the standard deviation is ± 3.7954.

II. Chromosome Analysis

A. *Introduction*

In addition to the studies of nuclear sex, more detailed cytological investigations of the chromosome complements of a number of athletes were also undertaken, using leucocyte culture.

The study of human chromosomes in diploid cells became successful after appropriate techniques for the observation of the chromosome complements of somatic cells had developed (Tjio and Levan, 1956; Hsu, 1952; Ford and Hamerton, 1956). The three principal steps leading to success were: (*1*) the improvement of the culture media in order to facilitate cell proliferation including the addition of antibiotics; (*2*) the addition of colchicine before the harvest of the cells in order to obtain metaphases; (*3*) the use of hypotonic solutions before fixation which makes the chromosomes spread on the slide.

Metaphases from peripheral blood leucocytes *in vitro* with a final spreading of the chromosomes have been obtained by drying smears in air (Moorhead *et al.*, 1960; Moorhead, 1962; Nowell, 1960).

These and other technical advances necessitated the seeking of an agreement concerning criteria for identification and classification of individual human chromosomes. The parameters for the construction of the ideogram and the karyotype, the individual characterization of most of the different chromosome pairs, and the rules for reporting chromosome anomalies have been established through several meetings of cytogeneticists in Denver (1960), London (1965), and Chicago (1966).

Extensive research work undertaken in many different places during the last decade has revealed a number of associations between clinical syndromes and structural as well as numerical chromosome aberrations, involving both autosomes and sex chromosomes. Most of this work which concerns severe abnormalities of the phenotype is not relevant for the present study.

However a search for chromosome anomalies in athletes is justified not only because chromosome anomalies in people, especially the relatives of patients, have been reported without noticeable defects on the phenotype, but also because variations in several quantitative traits associated with chromosome abnormalities and which could be inferred as enhancing sport abilities in some specialities also exist.

Chromosome anomalies in individuals without severe mental deficiency or congenital malformations or in families without physical or mental disabilities have been documented in the current cytological literature (Hungerford and Chandra, 1966; Kunze-Mühl *et al.*, 1970; Cohen *et al.*, 1970; Ridler *et al.*, 1970; Fredga and Hall, 1970).

Particularly interesting for the present study of chromosome anomalies in males and females are those affecting stature or causing dysplasias and sterility without necessarily weakening most of the physiological—mainly neuromuscular functions—required for Olympic performance. Since the early discovery of the most frequent sex chromosome anomalies 47/XXY and 45/XO associated with the Klinefelter and Turner syndromes and which result in high and low stature, respectively, a great number of aberrations involving the sex chromosomes and producing variable clinical syndromes have been described. These range from the normal

XX female and normal XY male to most complex numerical variations. (Ford *et al.,* 1959; Wiener and Sutherland, 1968; Stenchever and Macintyre, 1969; Berger, 1970; Cowling *et al.,* 1969; Grace and Harris, 1970; Jacobs *et al.,* 1959; Jacobs and Strong, 1959).

So far the sport abilities of almost normal phenotypes possessing chromosomal anomalies have not yet been specially investigated, but general observations show that they do not appear to be incapacitated. One could assume that some of them show anthropological traits relevant to certain sport specialities which would be potentially advantageous for the different sports. Two examples may illustrate this.

Patients with XYY karyotype are taller than average and having certain aggressive tendencies (at least some of them) might be very well suited for high jumps or basketball (Jacobs *et al.,* 1965; Berger, 1970; Nielsen, 1968).

Most extreme cases of pure gonadal dysgenesis with XY genotype and exceptionally female phenotype might perform better than normal females in some competitions. Also some true hermaphrodites who may for practical purposes be classified as men or women may be interesting (Harnden and Jacobs, 1961; Court-Brown *et al.,* 1964).

B. *Material and Methods*

Chromosomes were successfully analyzed from the cultures of peripheral blood taken from 227 male and 15 female athletes. In addition to association between karyotype details and sporting specialities, possible differences attributable to several racial groups were considered. (See Table 8.2.)

TABLE 8.2. DISTRIBUTION OF ATHLETES INVESTIGATED FOR KARYOTYPE

RACE			CHROMOSOME ANALYSIS	
			M	F
Caucasoid			129	12
	M	F		
(a) Negroid	40	2		
(b) Mulatto	8	0		
(c) Negroid + Mongoloid	0	0		
(d) Mestizo + Negroid	0	0		
Total NEGROID			48	2
Mongoloid			11	1
Mestizo			39	0
TOTAL			227	15

C. *Scoring*

Four observers worked as two pairs, two observers always independently scanning two slides from each subject. Counts of at least 15 cells per preparation were attempted and the first and any subsequent exceptional cell singled out for intensive investigation.

The preparations were systematically scanned under low-power, selecting for high-power examination only those cells which were satisfactory. Observers revealed their findings only when both subsamples of 15 cells each were completed.

The chromosomes in the mitotic cells were classified into the seven groups A to G, and the individual members of pairs: 1, 2, 3, 16 as well as the Y-chromosomes in males were distinguished. A normal male cell counted and analyzed in this way was recorded as:

	A 2/2/2
	B 4
	C 15
	D 6
2n = 46	E 2/4
	F 4
	G 4—Y

The results were put directly on cards with the karyotype photographs, prepared in the usual way.

D. *Results*

The chromosome analysis of the several hundred slides produced the remarkable finding that there exist extreme variations of the length of the Y-chromosome in male athletes of different races. A minority of individuals were carriers of very large Y-chromosomes, while in other individuals the dimensions of the Y-chromosome were quite small. In order to analyze this phenomenon, slides from 114 Caucasoid, 48 Negroid, 11 Mongoloid, and 34 Mestizo male athletes were selected for further study. For comparing the series of Y-chromosomes of these 207 male athletes in a more precise way a special parameter was selected, namely the ratio of the length of the Y-chromosome to that of the long arm of chromosome 2 (the Y/arm—2 ratio), Y_r, as follows:

$$Y_r = \frac{Y}{q}$$

where q is the length of the long arm of chromosome 2, and Y is the length of the Y-chromosome. A chromosome of the number 2 pair was selected at random. Starting at the centromere on the photograph the measurement was taken, by means of a compass, of the 2 chromatids of the long arm. Three to five cells per athlete were measured and the average value of the arm was obtained in millimeters. The Y-chromosome was measured from end to end in the same cells.

Individuals in whom the Y/arm—2 ratio departed more than one standard deviation from the population mean were selected in order to establish quantitative differences between racial groups and sport specialities.

However, as the sample contained individuals of four racial groups and different sport specialities a search was made for significant differences between racial groups or sport specialities.

For this, the values of the Y/arm—2 ratios were tabulated (see Table 8.3), computing the number of departures from the mean value of more than one standard deviation and using the formula:

$$D = \frac{\bar{Y}_r - Y_r}{\sigma}$$

where D represents the departures, in standard deviation units, and $\bar{Y}_r = (1/n)\ \Sigma Y_r$ and σ is the standard deviation of the sample, as per the following:

$$\sigma = \left[\frac{\Sigma Y_r^2 - [(\Sigma Y_r)^2/n]}{n-1}\right]^{\frac{1}{2}}$$

An analysis of the sample shows that 61 out of 207 Y-chromosomes (29.4%) fall outside the range of ±1 s.d. Among the 11 Mongoloid male athletes, 6 had larger Y-chromosomes than average and only one had a smaller than average Y-chromosome. These figures are suggestive of differences in lengths of the Y-chromosome from this racial group

TABLE 8.3. VARIATIONS OF THE Y/ARM-2 RATIO IN DIFFERENT RACIAL GROUPS

Race	No.	Variations of Y/arm-2 ratio in standard deviation units							
		≤ -2 s.d.		≤ -1 s.d.		1 s.d. ≤		2 s.d. ≤	
		n	%	n	%	n	%	n	%
Caucasoid	114	1	0.87	8	7.01	15	13.15	2	1.75
Negroid	48	0		5	10.41	10	20.83	3	6.25
Mongoloid	11	0		1	9.09	5	45.45	1	9.09
Mestizo	34	0		6	17.64	4	11.76	0	
Total	207	1		20		34		6	

compared with those from Caucasoids, Negroids, and Mestizos. As this has been previously reported (Cohen *et al.*, 1966), further investigations of this phenomenon should prove fruitful.

E. *Discussion*

Apart from possible increase of the length of the Y-chromosome in Mongoloids, no strong indications were found for systematic differences between other racial groups nor between athletes prominent in different sport specialities. However, as association between various phenotypes and the lengths of the Y-chromosomes have been described (Bishop *et al.*, 1962; Grifenberg, 1964; Makino et al., 1963; Tonomura *et al.*, 1963), investigations of the lengths of the Y-chromosome in various somatotypes predisposing for the different sport specialities might be promising, especially if larger samples might be available of Olympic athletes.

The absence of chromosome anomalies in our sample of Olympic athletes can be explained in several different ways. One could postulate that the sample was too small for the discovery of presumably rare abnormalities among the participants. On the other hand, the result may indicate that no such abnormalities confer advantages to the athletes and that even slight deviations from the chromosome norm are incompatible with Olympic performance.

Bibliography

Berger, R. (1970). The chromosome anomaly XYY. *Sci. Progr.* Paris **3424,** 35–36.

Barr, M. L. (1966). Correlation between sex chromatin patterns and sex chromosome complexes in man. *In* "The Sex Chromatin" (K. L. Moore, ed.), pp. 129–161. Saunders, Philadelphia and London.

Barr, M. L., and Bertram, E. G. (1949). A morphological distinction between neurones of the male and female and the behaviour of the nucleolar satellite during accelerated nuclear protein synthesis. *Nature* **163,** 676–677.

Barr, M. L., Bertram, E. G., and Lindsey, H. A. (1950). The morphology of the nerve cell nucleus according to sex. *Anat. Rec.* **107,** 283–297.

Commoner, B. (1964). Roles of deoxyribonucleic acid in inheritance. *Nature* **202,** 960–968.

Bishop, A., Blank, C. E., and Hunter, H. (1962). Heritable variation in length of the Y chromosome. *Lancet* (ii), 18–20.

Court-Brown, W. M., Harnden, D. G., Jacobs, P. A., Maclean, N., and Mantle, D. J. (1964). Abnormalities of the sex chromosome complement in man. *Med. Res. Coun. Spec. Recept. Ser.* **305.**

Court-Brown, W. M., Jacobs, P. A., and Price, W. H. (1968). Sex chromosome aneuploidy and criminal behaviour. *In* "Genetic and Environmental Influences on Behaviour" (J. M. Thoday and A. S. Parkes, eds.) pp. 180–193. Olver & Boyd, Edinburgh.

Cohen, M. M., Shaw, M. W. M., and Mac Cluer, J. W. (1966). Racial differences in the length of the Y chromosome. *Cytogenetics* **5**, 34–52.

Cowling, D. C., Rigo, S., Martin, F. I. R. (1969). The YY syndrome. *Med. J. Australia* **2**, 443–446.

Darlington, C. D., and La Cour, L. F. (1940). Nucleic acid starvation in *Trillium J. Genet.* **40**, 185–213.

Davidson, W. M. (1966). Sexual dimorphism in nuclei of polymorphonuclear leukocytes in various animals. *In* "The Sex Chromatin" (K. L. Moore, ed.) pp. 59–75. Saunders, Philadelphia and London.

Davidson, W. M., and Smith, D. R. (1954). A morphological sex difference in the polymorphonuclear neutrophil leucocytes. *Brit. Med. J.* **2**, 6–7.

De Grouchy, J., and Lamy, M. (1963). Mosaique X/X mosaique physigique chez la femme. *La Semaine des Hopitaux* (annales de Pediatrie) 99e année 17/4 pp 765/P. 163–768/P. 156.

De Grouchy, J., Jasso, N., De Gennes, J. L., Hennequet, F., Ribier, J., Frizal, J., and Lamy, M. (1963). Mosaiques interessant le chromosome X dans le syndrome de Turner. *La Semaine des Hopitaux* (annales de Pediatrie) 39e année *1714:* 769/P 157–774/P. 162.

Federman, D. D. (1967). "Abnormal Sexual Development." Saunders, Philadelphia and London.

Ford, C. E., and Hamerton, J. L. (1956). The chromosomes of man. *Nature* (*London*) **178**, 1020.

Ford, C. E., Jones, K. W., Polani, P. E., Almeida, J. C., and Briggs, J. H. (1959). A sex chromosome anomaly in a case of gonadal dysgenesis (Turner's Syndrome) *Lancet* **1**, 711–713.

Fredga, K., and Hall, B. (1970). A complex familial translocation involving chromosomes 5, 9 and 13. *Cytogenetics* **9**, 294–306.

German, J. L. (1964). The pattern of DNA synthesis in the chromosomes of human blood cells. *J. Cell Biol.* **20**, 37–55.

Goodman, R. M., Miller, F., and North, C. (1968). Chromosomes of tall men. *Lancet* (i), 1318.

Grace, H. J., and Harris, E. (1970). Familial of an abnormal Y chromosome. *S. Afr. Med. J.* **44**, 643–694.

Grifenberg, U. (1964). Size variation and orientation of the Y chromosome. *Chromosoma* **15**, 618–629.

Grumbach, M. M., Morishima, A., and Taylor, J. H. (1963). Human sex chromosome abnormalities in relation to DNA replication and heterochromatinization. *Proc. Nat. Acad. Sci.* **49**, 581–589.

Harnden, D. G., and Jacobs, P. A. (1961). Cytogenetics of abnormal sexual development in man. *Brit. Med. Bull.* **17**, 206–212.

Heitz, E. (1928). Das Heterochromatin der Moose. I. *Jahrb. wiss. Bot.* **69**, 762–818.

Heitz, E. (1933). Die somatische Heteropyknose bei Drosophila melanogaster und ihre genetische Bedeutung. *Z. Zellforsch. mirosk. Anat.* **20**, 237–287.

Holhn, H., Reinween, H., and Engel, W. (1970). Genetic studies on a minute centric fragment transmited through three generations, *Cytogenetics* **9**, 186–198.

Hsu, T. C. (1952). Mammalian chromosomes in vitro. I. The karyotype of man. *J. Hered.* **43**, 167–172.

Hungerford, D. A., and Chandra, H. S. (1966). Further studies of a kindred having an aberrant autosome (13–15) without apparent phenotypic effect. *J. Genet.* **59**, 239–244.

Jacobs, P. A., and Strong, J. A. (1959). A case of human intersexuality having a possible XXY sex-determining mechanism. *Nature* **183**, 302–303.

Jacobs, P. A., Baikie, A. G., Court-Brown, W. M., MacGregor, T. N., Maclean, N., and Harnden, D. G. (1959). Evidence for the existence of the human "Superfemale." *Lancet* (ii) 423–425.

Jacobs, P. A., Brunton, M., Melvile, M. M., Brittain, R. P., and McClemont, W. F. (1965). Aggressive behaviour mental subnormality and the XYY male. *Nature,* **208,** 1351–1352.

Klinger, H. P., and Schwarzacher, H. G. (1962). XX/XXY and sex chromatin possitive cell distribution in a 60 mm human fetus. *Cytogenetics* **1,** 266–290.

Kunze-Muhl, E. P., Fischer, E., and Golob, F. (1969). Cytogenetic studies of 7 relatives of whom 2 carried a balanced g/g translocation. *Wiener Klin.* Wochensch. **81,** 637–638.

Kunze-Muhl, E. P., Fischer, E., and Golob, F. (1970). Familiar appearance of a C-chromosome aberration. *Humangenetik* **9,** 91–96.

Lyon, M. F. (1961). Gene action in the X-chromosome of the mouse (*Mus musculus* L.) *Nature* **190,** 372–373.

Lyon, M. F. (1966). Sex chromatin and gene action in the X-chromosome of mamals. *In* "The Sex Chromatin" (K. L. Moore, ed.) pp. 370–386. Saunders, Philadelphia and London.

Maclean, N. (1966). Sex chromatin surveys of newborn babies. In "The Sex Chromatin" (K. L. Moore, ed.) pp. 202–210. Saunders, Philadelphia and London.

Maclean, N., Mitchell, J. M., Harnden, D. C., Williams, J., Jacobs, P. A., Buckton, K. A., Baikie, A. G., Court Brown, W. M., McBride, J. A., Strong, J. A., Close, H. G., and Jones, D. C. (1962). A survey of sex-chromosome abnormalities among 4514 mental defectives. *Lancet* (i), 293–296.

Maclean, N., Court Brown, W. M., Jacobs, P. A., Mantle, D. J., and Strong, J. A. (1968). A survey of sex chromatin abnormalities in mental hospitals. *J. Med. Genet.* **5,** 165–172.

Makino, S., Sasaki, M. S., Yamada, K. M., and Kuji, T. (1963). A long Y chromosome in man. *Chromosoma* **14,** 154–161.

Marberger, E., Boccabella, R. A., and Nelson, W. O. (1955). Oral smear as a method of chromosomal sex detection. *Proc. Soc. Exp. Biol. Med.* **89,** 488–489.

Mather, K. (1944). The genetical activity of heterochromatin. *Proc. Roy. Soc. B.* **132,** 309–332.

Mittwoch, U. (1967). "Sex Chromosomes." Academic Press, New York and London.

Moore, K. L., and Barr, M. L. (1955). Smears from the oral mucosa in the detection of chromosomal sex. *Lancet* **2,** 57–58.

Moore, K. L., Graham, M. A., and Barr, M. L. (1953). The detection of chromosomal sex in hermaphrodites from a skin biopsy. *Surg. Gynec. Obstet.* **96,** 641–648.

Moorhead, P. S. (1962). The blood technique of human chromosomes. *Symp. Mamm. Tissue Culture Cytol.* Sao Paulo 1962. Macmillan (Pergamon), New York.

Moorhead, P. S., Nowell, P. S., Mellman, W. J., Battips, D. M., and Hungerford, D. A. (1960). Chromosome preparations of leukocytes cultured from human peripheral blood. *Exptl. Cell Res.* **20,** 613–616.

Muller, H. J., and Painter, T. S. (1932). The differentiation of the sex chromosomes of *Drosophilia* into genetically active and inert regions. *Z. Induktive Abstammungs-Vererbungslehre* **62,** 316–365.

Nielsen, J. (1968). The XYY chromosomal constitution in criminal psycopaths. *Lancet* (ii), 576.

Nowell, P. C. (1960). Phytohemagglutinin: an initiator of mitosis in cultures of normal human leukocytes. *Cancer Res.* **20,** 462–466.

Pearse, A. G. E. (1968). "Histochemistry," (3rd ed.) Vol. 1. Churchill, London.

Ricci, N., Dallapiccola, B., Ventimiglia, B., Tiepolo, L., and Fraccaro, M. (1968). 48, XXXX/49, XXXXX mosaic asynchronies among the latereplicating X chromosomes. *Cytogenetics* **7**, 249–259.

Ridler, M. A. C., Berg, J. M., Pendrey, W. J., Saldana, P., and Timothy, J. A. D. (1970). Familial occurrence of a small, supernumerary metacentric chromosome in phenotypically normal women. *J. Med. Genet.* **7**, 148–152.

Stenchever, M. A., and Macintyre, M. N. (1969). A normal XYY man. *Lancet* (i), 680.

Tjio, J. H., and Levan, A. (1956). The chromosome number of man. *Hereditas* **2**, 1–6.

Tonomura, A., and Ono, H. (1963). Variation in length of the Y chromosome in man. *Jap. Nat. Inst. Gen. Report* **14**, 131.

Wiener, S., and Sutherland, G. (1968). A normal XYY man. *Lancet* (ii), 1352.

9

Retrospect

When the ancient Greeks gathered at Olympia, they came not only to observe the performances of their athletes, but also those of their poets and musicians. In our more fragmented civilization these diverse manifestations of human genius are unfortunately often not admired by the same people, in spite of the fact that they have many features in common. These common features include their uniqueness and complexity and, biologically speaking, their dependence on hereditary endowment, general environmental conditions, and special training.

The study of athletes is, thus, in line with the study of any other occupational group, be they artisans or professional men, and their extreme performances elicit such questions as: "What makes a good runner?" Or, as seen from the individual's point of view: "At what sport am I likely to excel?" Insofar as the unique personal features of every athlete are concerned, this question eludes scientific analysis, but enough is left of common features, both in respect to high general performance and to excellence in the various sport disciplines (specialities), to warrant a systematic investigation of the biological foundations on which athletic success must be built.

I. Biological Approach to Sport

Athletes are both born and made; and there is little doubt that only a rare combination of genetic endowment, generally good environmental conditions, and very special training will, in today's advanced stage of competition, produce performances of Olympic standard. The present study is concerned with those factors, genetic as well as environmental, which predispose a person to high athletic achievement. Not considered in these studies are the methods of specialized training which turn a suitable candidate into a top athlete and which are a facet of the field of sport medicine. We have also excluded another branch of sport medicine—sport physiology—which is widely studied by other people.

The investigators responsible for this study were either physical anthropologists or geneticists, and they brought to this study their various approaches to the subject. The anthropologists, drawing on previous experience, proceeded to type and measure the athletes and then correlated their findings with the particular sports of the athletes. These anthropometric–anthroposcopic characteristics are of course themselves products of the genetic constitutions and the environmental conditions in which the athletes grew up and were trained. While these interacting factors make it difficult to answer general questions, there is a great advantage in the anthropologists' approach in that it tries to correlate such obviously connected entities as body build, size, or age, with quantitative characteristics, namely the special performances, that are so directly dependent upon these entities.

With respect to the genetic studies, the cause and effect relationship is by no means clear. There has been no good correlation found between any specific genes and any specific athletic achievement. To be sure there is no a priori reason to assume any particular connection between any allele and any special achievement in a sport discipline. Yet some connections between sports ability and genetic endowment must be presumed to exist. The genetic studies detailed in this book have searched for just such connections.

The genetics of quantitative characteristics, such as stature or intelligence, have been approached in two quite different ways, which we may call the study of the normal and the study of the pathological. A number of single genes have been found which render the individual a particular type of dwarf, giant, or idiot, as the case may be. These genes segregate clearly, in simple hereditary fashion, in the families in which they occur. In our study of the athletes we have been, with one exception, concerned with such a situation only in a negative way. This negative approach has sought to determine whether a particular genotype (e.g., nontasting of PTC or sickling of red blood cells) puts an athlete at a disadvantage. Otherwise we have dealt with the tail-ends of the more or less normal distributions where many alleles at many loci are implicated in the formation of the particular capacities for the various athletic specialities. Thus while one "wrong" gene may ruin the chances of an individual to become a good athlete, or for that matter, an astronaut, top mathematician, musician, painter, etc., a multitude of favorable genes must combine to endow the individual with potential excellence in the field. No single gene has been discovered or is likely to be found which makes an individual an Olympic athlete.

Another aspect of the study of genetic components in relationship to sports ability is a comparison of the sports achievements of the major human races. It is important to note that the various human populations do contribute different proportions of competitors in the various types of Olympic events. However, we cannot as yet explain why this is so. The excellence in sprinting or jumping of Negroids could for instance be based on their anatomical measurements, on their social motivation, or very possibly on other factors. This question remains a subject for further study. On the whole, it is found that the athletes come from the densely populated urban centers, most divorced from the relatively simple lives of tribal hunters, gatherers, or farmers. This would appear to be unexpected since the more active rural environment would seem to provide much more opportunity for the "basic" training of an athlete than would the sedentary urban existence. However, this may be just a statistical phenomenon caused by the numerical preponderance of the urban element in some participating nations and by the greater urban opportunities for becoming an athlete.

The practical applications of our results, though modest, are of two kinds. On the one hand they may enable the selector of sport candidates to better choose the correct ones for training in the various events. The second practical application of our results is the help that this genetic and anthropometric information may provide for the intending sportsman and sportswoman, as well as their medical and educational advisers, in choosing the right specialities.

II. Implications of These Studies for Human Biology

The isolation and identification of specific athletic traits adds yet another dimension to the study of human populations, just as do studies of intelligence, musical ability, or mathematical talent. Once identified, each trait (be it sprinting ability, running endurance, or the "explosive" capacities of jumpers and throwers) can be correlated with any of the numerous factors that may affect athletic potential, such as inbreeding, assortative mating, morbidity, fertility, and the like. It is also true that the particular abilities of sportsmen and sportswomen can be correlated with one another, since these various abilities are quite frequently jointly present in one individual. In addition, it may prove valuable to see if any correlation exists between specific athletic traits and other individual traits that are not commonly considered as contributory to the performance of athletes.

Certain difficulties that occur in defining our traits in a sample drawn from many ethnic groups can, if surmountable, be turned into new opportunities. An outstanding athlete, like a great musician or painter, can as a rule be recognized independently of his racial or cultural origins. In like fashion, groups of such outstanding individuals, drawn from a multitude of people, must have common properties, independent of their various backgrounds. Nevertheless there are considerable difficulties in drawing valid conclusions, mostly of a statistical nature. Top athletes do not, for instance, with respect to the traits which make them excel, form a normal population. Rather they are part of the tail-end of such a distribution. Therefore, some statistics of extremes rather than the more usual statistics of normal distribution must be applied to the data involving these traits. Fortunately, such approaches have been developed for industrial purposes and are available for these studies.

A second difficulty arises when choosing a control population for any particular comparison. However, this problem may also be considered as a challenge rather than a hindrance, whose solution will result in a better understanding of the bases of human abilities.

Insofar as the performance of men and women differ in comparable events, they can contribute to the assessment of the physiologically and genetically determined differences between the sexes. These differences are probably of varying magnitude in the several human races, indicating different solutions to a most important problem of adaptation to a similar environment. The study of various "intersexes" among athletes will undoubtedly contribute a good deal to our understanding of human achievements. Thus we can see the relevance of these studies of Olympic athletes to other areas of human biology.

Index

A

ABO blood groups, *see* Red blood cell antigens
Acid phosphatase, *see* Red blood cell enzymes
Age, 12, 28, 32, 35–36, 38, 40–41, 44–46, 48–49, 51–52, 56, 82–128, 130–146
Anthropometric instruments, 76
Anthropometric measurements, 76–82
Analysis
  by race, 61–68
  between sports, 50–61
  within sports, 27–50
Arm length, 54, 59–60, 82–128, 130–146
  definition of, 81
Athletes
  classification of, 20–22
  numbers of, 20–22, 129–130
  racial categories, 21–22, 129–133
Athletic ability of parents of athletes, *see* Family studies
Athletic ability of parents of control group, *see* Family studies

B

Barr body, *see* Sex chromatin analysis
Basketball, 39–40, 52–55, 64–65, 103, 105, 127, 129, 141
Biacromial breadth, 53, 59, 82–128, 130–146
  definition of, 78
Biiliocristal breadth, 54, 59, 82–128, 130–146
  definition of, 79
Birth order of athletes, *see* Family studies
Blood groups, *see* Red blood cell antigens
Blood plasma proteins, 165–169, 178–181
  group specific components (Gc), 181, 182
  haptoglobin, 179, 180
  transferrin, 180, 181
Body size, 30, 34, 36–41, 43, 45, 47–48, 50, 58–60, 62–73
Boxing, 45–47, 55, 66–67, 114, 119–123, 129, 144–145

C

Canoeing
  men, 41, 52–55, 103, 108–110, 127, 129
  women, 41–42, 52–55, 101, 110, 128, 130

Caucasoid, *see also* Race
men athletes, 129, 130
reference men, 134
reference women, 135
women athletes, 130, 132
Chromosome analysis, 219–224
history of, 219, 220
results of, 221–224
technique and evaluation of, 220–222
Components, *see* Somatotype
Cycling, 41–43, 52–55, 65, 103, 111–113, 127, 129, 142
Cytogenetics, *see* Sex chromatin analysis; Chromosome analysis

D

Dermatoglyphics, *see* Finger prints; Palm prints
Diving
men, 38, 52–55, 102–103, 127, 219
women, 38–39, 52–55, 101, 103, 128, 130
Drumstick, *see* Sex chromatin analysis

E

Ectomorphy, 15, 53, 57, 75, 82–128, 130–146
Endomorphy, 15, 52, 56, 74, 82–128, 130–146
Event, 20–21, *see also* names of specific sports groups
numbers in, 129–130

F

Family studies, 147–155
history of, 147, 148
results obtained in, 149–155
technique used in, 148, 149
Field events (men), 31–34, 55, 63, 86–90
Finger prints, 189–192, 198–206
experimental technique, 198, 199
inheritance of, 197
pattern index intensity, 190, 200–202, 206
pattern type, 190, 191, 204, 205
ridge count, 191, 192, 198, 206

G

Glucose-6-phosphate dehydrogenase, *see* Red blood cell enzymes
Group specific components (Gc), *see* Blood plasma proteins
Gymnastics
men, 43, 52–55, 65–66, 113–114, 127, 129, 143
women, 43, 52–55, 66, 101, 114, 128, 130, 143

H

Haptoglobin, *see* Blood plasma proteins
Height, maximum stretch, 53, 58, 69, 82–128, 130–146
definition of, 77
Height–weight ratio, 15, 53, 75, 82–128, 130–146
Hemoglobin A, C, S, *see* Red blood cell proteins
Hip breadth, *see* Biiliocristal breadth

L

Leg length, 54, 60, 82–128, 130–146
definition of, 81

M

Medalists, 20
Mesomorphy, 15, 52, 56–57, 74, 82–128, 130–146
Mestizo, *see also* Race
men athletes, 129, 131
reference men, 134
reference women, 135
women athletes, 130, 133
MN blood groups, *see* Red blood cell antigens
Modern pentathlon, 39, 52–55, 103–104, 127, 129
Mongoloid, *see also* Race
men athletes, 129, 131
women athletes, 130

N

Negroid, *see also* Race
  men athletes, 129, 132
  women athletes, 130, 133

O

Olympic athletes
  previous studies of, 10–14, 71

P

Palm prints, 189, 190, 192–198
  axial triradius, 194
  digital triradii, 193, 194
  experimental technique, 198, 199
  inheritance of, 197
  main-line index, 195, 196, 203, 208
  main lines, 194–196, 207
  maximal *atd* angle, 194, 196–198, 209
Phenylthiourea sensitivity, 157–164
  discovery of, 158
  genetics of, 158
  racial groups and, 161, 162
  sex of athlete and, 161, 162
  sport specialities and, 161–163
  test procedure, 159, 160
Physical anthropology
  methods and procedures, 14–24
Physique, *see* Somatotype
PTC sensitivity, *see* Phenylthiourea sensitivity

R

Race
  analysis by, 61–68, 71–73, 130–146
  categories, 21–22
  numbers in categories, 129–130
Red blood cell antigens, 165–172
  ABO, 169–171
  MN, 171, 172
  Rh–Hr, 172, 173
Red blood cell enzymes, 165–169, 172–178
  acid phosphatase, 175–178
  glucose-6-phosphate dehydrogenase, 174, 175
Red blood cell proteins, 165–169, 178, 181–184
  hemoglobin A, 181–184
  hemoglobin C, 181–184
  hemoglobin S, 181–184
Reference
  men, 22, 49, 52–55, 67, 126, 135, 146
  women, 22, 49, 52–55, 68, 127, 134, 146
Rh–Hr blood groups, *see* Red blood cell antigens
Rowing, 40, 52–55, 65, 103, 105–108, 127, 129, 142

S

Sex chromatin analysis, 211–219
  history of, 211, 212
  results of, 219
  technique and evaluation of, 212–218
Sex groups, 20–21
  by sport, event, and race, 129–130
Shoulder breadth, *see* Biacromial breadth
Skinfolds, 22, 52, 74, 80–81
  sum of three, 22, 82–128, 130–146
Somatochart, 17–18, 55
  distribution, 57–58
Somatometry, 14, 20
Somatotype, 14–19, 28, 30, 32, 35–46, 48–50, 55–56, 62–73
  anthropometric, 15–16, 74–75
  categories, 18–19
  components, 15, 74–75
  definition, 15
  distribution, 17, 55, 57–58
  frequencies, 86, 94, 98, 101
  photograph, 16
  rating form, 16, 74
Somatotyping
  Cureton's method of, 11
  Heath–Carter method of, 15–16
  instructions for, 74–75
Sports, *see also* name of sport
  analysis between, 50–61, 127–128
  analysis within, 27–50, 82–127
  groupings, 20–21
  numbers in, 20–21, 129–130
Statistical procedures, 23–24
Stature, *see* Height, Body size

Swimming
men, 36–37, 52–55, 64, 95–98, 127, 129, 140
women, 37–38, 52–55, 64, 99–102, 128, 130, 141

T

Total leg length, right, *see* Leg length
Track
men, 28–31, 52–55, 62–63, 82–86, 127, 129, 136–138
Track and field
men, 28–34, 52–55, 62–63, 82–90, 127, 129, 136–138
women, 34–36, 52–55, 63–64, 90–94, 128, 130, 139–140
Transferrin, *see* Blood plasma proteins
Trunk length, 54, 60, 82–128, 130–146
definition of, 81

U

Upper plus lower arm length, right, *see* Arm length

V

Variables
analyzed, 22
choice of, 22–23
measured, 76–82

W

Water polo, 39, 52–55, 103–104, 127, 129
Weight, 53, 58–59, 76, 82–128, 130–146
Weight lifting, 47–48, 55, 67, 114, 124–126, 129, 145
Wrestling, 43–45, 55, 66, 114–119, 129, 144

A 4
B 5
C 6
D 7
E 8
F 9
G 0
H 1
I 2
J 3